TK
7867
.M44
1978

Loudoun, Campbell.
　Handbook for electronic
circuit design.

HANDBOOK FOR
Electronic Circuit
Design

HANDBOOK FOR
Electronic Circuit Design

Campbell Loudoun

RESTON PUBLISHING COMPANY, INC.
A *Prentice-Hall Company*
Reston, Virginia

Library of Congress Cataloging in Publication Data
Loudoun, Campbell
  Handbook for electronic circuit design.

  Includes index.
  1. Electronic circuit design—Handbooks, manuals, etc. I. Title.
  TK7867.M44   621.3815'3   77-23981
  ISBN 0-87909-334-X

© 1978 by Reston Publishing Company, Inc.
A Prentice-Hall Company
Reston, Virginia 22090

All rights reserved. No part of this book may be reproduced in any way, or by any means, without permission in writing from the publisher.

10 9 8 7 6 5 4 3 2 1

PRINTED IN THE UNITED STATES OF AMERICA

Contents

Preface   ix

Chapter 1
Introduction to Design Principles and Parameters   1

    1-1   COMPONENT AND DEVICE TOLERANCES   1
    1-2   WORST-CASE VALUES   3
    1-3   GENERALIZED RESISTIVE NETWORK   3
    1-4   HOMOLOGOUS AND HETEROLOGOUS TOLERANCES   4
    1-5   BASIC DEVICE TOLERANCES   5
    1-6   NOTES ON NEGATIVE FEEDBACK   7
    1-7   POWER DISSIPATION PRINCIPLES   9
    1-8   LAW OF NORMAL DISTRIBUTION   11
    1-9   NOTES ON MATHEMATICAL MODELS   12
    1-10   COMPUTER-AIDED DESIGN   15
    1-11   PITFALLS IN DESIGN PROCEDURES   16
    1-12   HIDDEN VARIABLES   18
    1-13   UNDERWRITERS' LABORATORIES REQUIREMENTS   20
    1-14   ENVIRONMENTAL TESTING   21
    1-15   BURN-IN DESIGN TEST PROCEDURES   22
    1-16   LIFE TESTS   22

vi · Contents

Chapter 2
Elements of Electronic Circuit Design 27
    2-1 RC FILTER PRINCIPLES 27
    2-2 NORMALIZATION TECHNIQUE OF TOLERANCE ANALYSIS 37
    2-3 OPERATION OF RESISTORS AT HIGH FREQUENCIES 38
    2-4 CASCADED RC CIRCUITRY 39
    2-5 BASIC GRAPHICAL SOLUTIONS 41
    2-6 BASIC RC BANDPASS FILTER ACTION 43
    2-7 RC BAND ELIMINATION FILTER 45
    2-8 TRANSIENT RESPONSE OF MULTISECTION RC FILTERS 47
    2-9 MONTE CARLO METHOD 49
    2-10 CIRCUIT DESIGN BREADBOARD 49

Chapter 3
Fundamentals of Wave Filter Design 51
    3-1 GENERAL CONSIDERATIONS 51
    3-2 MULTISECTION LC WAVE FILTER 56
    3-3 ELEMENTS OF ACTIVE FILTERS 68
    3-4 INCOMING INSPECTION PROCEDURES 76

Chapter 4
Basic Audio Amplifier Design 77
    4-1 GENERAL CONSIDERATIONS 77
    4-2 BASIC AMPLIFIER CHARACTERISTICS 79
    4-3 INPUT RESISTANCE VERSUS LOAD RESISTANCE 82
    4-4 OUTPUT RESISTANCE VERSUS GENERATOR RESISTANCE 83
    4-5 VOLTAGE AND CURRENT AMPLIFICATION 83
    4-6 BASIC FET AMPLIFIER CHARACTERISTICS 85
    4-7 LOAD LINES 86
    4-8 ELEMENTS OF NEGATIVE FEEDBACK 91
    4-9 ACTIVE TONE CONTROL 101
    4-10 MULTISTAGE NEGATIVE-FEEDBACK HAZARD 105

Chapter 5
Amplifier Bias Stabilization 109
    5-1 GENERAL CONSIDERATIONS 109
    5-2 RESISTOR STABILIZING CIRCUITS 113
    5-3 VOLTAGE STABILITY FACTOR 118
    5-4 THERMISTOR STABILIZING CIRCUITS 119
    5-5 DIODE STABILIZING CIRCUITS 122
    5-6 TRANSISTOR STABILIZED AMPLIFIER CIRCUITS 126

5-7 BREAKDOWN DIODE VOLTAGE REGULATOR 130
5-8 INNOVATIVE DESIGN 134

Chapter 6
Fundamentals of Radio-Frequency Circuit Design 135
6-1 GENERAL CONSIDERATIONS 135
6-2 BASIC RESONANT CIRCUIT DESIGN 137
6-3 TRANSISTOR AND COUPLING NETWORK IMPEDANCES 142
6-4 TRANSFORMER COUPLING WITH TUNED PRIMARY 142
6-5 INTERSTAGE COUPLING WITH DOUBLE-TUNED NETWORKS 146
6-6 NEUTRALIZATION AND UNILATERALIZATION 152
6-7 CALCULATOR AIDED DESIGN 157
6-8 SCREEN ROOM ADVANTAGES IN RF DESIGN 160
6-9 GOOD DESIGN PRACTICES 160

Chapter 7
Principles of Oscillator Circuit Design 161
7-1 GENERAL CONSIDERATIONS 161
7-2 TRANSISTOR IMPEDANCE CONSIDERATIONS 162
7-3 INITIAL AND SUSTAINED OSCILLATION 163
7-4 FREQUENCY STABILITY 165
7-5 BASIC TRANSISTOR OSCILLATOR CONFIGURATIONS 168
7-6 COLPITTS AND CLAPP OSCILLATOR OPERATION 174
7-7 HARTLEY OSCILLATOR OPERATION 177
7-8 CRYSTAL OSCILLATORS 180

Chapter 8
Nonsinusoidal Oscillator Circuit Design Principles 186
8-1 GENERAL CONSIDERATIONS 186
8-2 BISTABLE MULTIVIBRATOR OPERATION 190
8-3 NONSATURATING BISTABLE MULTIVIBRATOR 197
8-4 MONOSTABLE MULTIVIBRATOR 197
8-5 SCHMITT TRIGGER (SQUARING) CIRCUIT OPERATION 199
8-6 TRIGGERED BLOCKING OSCILLATOR OPERATION 201
8-7 SWITCHING-CIRCUIT CHARACTERISTICS 206
8-8 UJT RELAXATION OSCILLATOR DESIGN 210

Chapter 9
AF and RF Power Amplifier Fundamentals 212
9-1 GENERAL CONSIDERATIONS 212
9-2 THERMAL RUNAWAY HAZARDS 212

9-3 COMPLEMENTARY SYMMETRY AMPLIFIER OPERATION 213
9-4 RF POWER AMPLIFIERS 227

Chapter 10
Debugging Logic Systems 231
10-1 GENERAL CONSIDERATIONS 231
10-2 FUNCTIONAL ANALYSIS OF NATIONAL IMP MICROPROCESSOR SYSTEMS 243

Appendix I:
Color Codes 253

Appendix II:
Characteristics of Series- and Parallel-resonant Circuits 259

Appendix III:
Graphical Solution for Inductive and Capacitive Reactances in Parallel 261

Appendix IV:
Logic Gates and DoMorgan's Equivalents 262

Appendix V:
Integrated Injection Logic (I²L) 263

Appendix VI:
Standard Potentiometer Resistance Tapers 266

Appendix VII:
Resistance RMA Preferred Values 267

Index 269

Preface

Electronic circuit design is concerned with the planning and detailing of configurations to meet particular performance specifications within the framework of reproducible design in large-scale production. In implementation, all component and device tolerances and ratings are analyzed within the context of these specifications to ensure that the design will not "bomb out" in its production phase. Since most electrical tolerances have a finite life expectancy, the circuit designer evaluates reliability requirements and allows a corresponding margin for component and device deterioration.

The keystone of circuit design procedure is worst-case analysis. This is a design technique wherein circuit performance is determined for the case in which all components and devices have simultaneously assumed their most unfavorable tolerance values, in which the supply voltage is minimum, and in which environmental conditions (temperature, for example) are most adverse. Worst-case analysis may also be extended to include deteriorated performance owing to defined aging processes. Conservative derating factors applied to design-center or bogie values are necessarily related to worst-case conditions. Since uncompromising design measures are generally unacceptable from the viewpoint of production costs, the circuit designer is confronted with the responsibility for making a judicious choice of evils; he is forced to accept certain calculated risks.

In various projects, the circuit designer must make environmental tests of his prototype model under conditions such as heat, shock, vibration, reduced atmospheric pressure, moisture, dust, radiation, lightning, and so on. Accelerated life tests are often advisable. Quality-control procedures must often be established to avoid the hazard of ruinous deviations or faults in production runs. Incoming inspection procedures may be set up in various situations to ensure that components, devices, and hardware are within specified tolerances before they go into production. Underwriters' Laboratories approval, if applicable, must be sought before the prototype goes into production. Many municipalities require UL approval for all products sold in their jurisdictional area.

An introduction to design principles and parameters is presented in the first chapter of this text. Common pitfalls and hidden variables are pointed out. The second chapter discusses the elements of electronic circuit design, with tolerance analyses of RC filter circuitry. Wave-filter design is treated in the third chapter for constant-K and composite configurations, with an introduction to op-amp active filter design parameters. In the fourth chapter, basic audio amplifier design considerations are explained; prior introduction to negative-feedback principles is developed, with notes on hazards of multistage feedback. Transistor worst-case factors are discussed.

Amplifier bias stabilization requirements are described in the fifth chapter, with evaluations of current-stability and voltage-stability factors. Fundamental principles of radio-frequency circuit design are presented in the sixth chapter. Impedance-matching methods are illustrated, with worst-case evaluations. Practical means of contending with production tolerances are explained. Good design practices are noted. The seventh chapter covers the principles of oscillator circuit design. Dropout hazards are analyzed, and design safeguards are noted. Temperature-compensating capacitor function is illustrated. In Chapter 8, nonsinusoidal oscillator circuit-design principles are developed. Switching-time factors are presented for both saturated and unsaturated modes of operation. Worst-case factors are pointed out, with the device and circuit-action conditions that are involved.

Chapter 9 treats AF and RF power-amplifier fundamentals, with stress on design parameters. Hazards of thermal runaway are explained, and basic precautionary measures are noted. Good practices are discussed in regard to parts layout and grounding arrangements in RF power-amplifier circuitry. Parasitics are almost always a design problem, and control methods are described. Hazards of shunt feed in solid-state power amplifiers are pointed out and analyzed. The tenth chapter is concerned with debugging of digital logic systems during the design phase. Because of the mushrooming popularity of microprocessor systems, this area is stressed in the practical examples that are cited. Operation of logic state analyzers and interpretation of test data are emphasized. However, the relation of the time domain to the data domain is not ignored. Map displays are illustrated, and their application to debugging procedures is noted.

It is assumed that the student has completed courses in basic electricity, electronics, and digital logic theory. In turn, this text has been prepared to fill the gap between academic circuit theory and circuit design-engineering techniques. It will be found equally valuable in structured classroom instruction, in the laboratory or shop, and in the home for self-instruction. The author is indebted to his associates, who have made many constructive criticisms and helpful suggestions. He is also grateful to the Hewlett-Packard organization, in particular, for permission to reproduce highly informative application notes. It is appropriate that this text be dedicated as a teaching tool to the instructors and students of our junior colleges and community colleges, technical institutes, and vocational schools.

CAMPBELL LOUDOUN

CHAPTER 1

Introduction to Design Principles and Parameters

1–1 COMPONENT AND DEVICE TOLERANCES

Components and devices have exact values and ideal characteristics only in theory. In practice, every fabricated unit is subject to tolerances. As an illustration, a metal-film resistor may have a rated resistance value of 10,000 ohms, with a tolerance of ±1 percent. This tolerance rating denotes that the actual value of a resistor chosen at random in a production lot could range in value from 9900 ohms to 10,100 ohms. The circuit designer is ordinarily justified in the assumption that all resistors in a production lot will be within rated tolerance. However, it cannot be assumed that all of the resistors in the production lot will remain in tolerance after undergoing production operations. After the resistor leads have been clipped and bent, subjected to torsion, soldered into a circuit board, and cycled numerous times at various temperatures, it is not necessarily true that all resistors in the production lot will then be within rated tolerance. Accordingly, the circuit designer is often required to establish production test procedures that will indicate whether resistors (and other components) have remained within rated tolerances.

When resistors with the same tolerance rating are connected in *series,* the tolerance on the series combination is the same as the tolerance on an individual resistor. For example, if three 10,000-ohm composition resistors are connected in series, and each resistor has a tolerance of ±5 percent, the series combination could range in value from 28,500 ohms to 31,500 ohms. Or, as depicted in Fig. 1–1, the

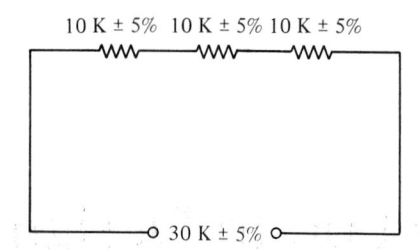

Figure 1–1 Example of tolerances in a series string of resistors.

resistance value of the combination is expressed as 30,000 ohms ±5 percent. This principle is equally true for series-connected resistors that have unequal nominal values with the same tolerance rating. As an illustration, if a series combination comprises a 5000-ohm resistor, 10,000-ohm resistor, and a 15,000-ohm resistor, each with a tolerance of ±10 percent, the series combination will have a nominal value of 30,000 ohms with a tolerance of ±10 percent.

However, if series-connected resistors have equal nominal (bogie) values, with unequal tolerances, the tolerance on the combination is equal to the mean (average) value of the individual tolerances. Thus, if a 10,000-ohm ±10 percent resistor is connected in series with a 10,000-ohm ±20 percent resistor, the tolerance of the 20,000-ohm combination is ±15 percent. Again, if a 1000-ohm ±5 percent resistor is connected in series with a 1000-ohm ±10 percent resistor, and a 1000-ohm ±15 percent resistor, their combination has a nominal value of 3000 ohms with a ±10 percent tolerance. On the other hand, if a pair of series-connected resistors have unequal bogie values and unequal tolerance ratings, the tolerance on their series combination will *not* be equal to their mean tolerance, although their series-combination tolerance will fall somewhere in the range from the smaller individual tolerance to the larger.

When resistors with the same tolerance rating are connected in parallel, the tolerance on their parallel combination is the same as on an individual resistor. Again, when resistors with the same tolerance rating are connected in series-parallel, the tolerance on their series-parallel combination is the same as on an individual resistor. However, when resistors with different tolerance ratings are connected in parallel, the tolerance on their parallel combination will *not* be equal to their mean tolerance. It follows that when resistors with different tolerance ratings are connected in series-parallel, that the tolerance on their series-parallel combination will *not* be equal to their mean tolerance, unless all of the parallel-connected resistors happen to have the same tolerance rating.

1–2 WORST-CASE VALUES

Worst-case circuit analysis is an evaluation procedure that is used to determine the worst possible effect on the output parameters owing to changes in the values of circuit elements. The circuit elements are set to the values within their anticipated ranges that produce the maximum detrimental changes in the output. *Worst-case design* is an extremely conservative design approach in which the circuit is designed to function normally, even though all of the component values have simultaneously assumed the worst possible condition that can be caused by initial tolerance, aging, and so on. Worst-case techniques are also applied to obtain conservative derating of transient and speed specifications.

Consider three 10,000-ohm ±1 percent resistors connected in series. Their series combination could range in value from 29,700 ohms (with all tolerances on the low side), to 30,300 ohms (with all tolerances on the high side). Since these values represent the worst possible effect on the combination, these range limits are called *worst-case values*. Of course, this is a theoretical treatment of a worst-case problem. It follows from previous discussion that, after the resistor leads have been clipped and bent, soldered into a circuit board, and cycled numerous times at various temperatures, electrical measurements would possibly show that the calculated worst-case values must be extended to a higher limit, or to a lower limit, or both.

1–3 GENERALIZED RESISTIVE NETWORK

Insofar as input/output characteristics are concerned, any linear resistive network can be reduced to a T section, or pad. The equivalence of a network and a T pad is based on input-resistance measurements with the output terminals open-circuited or short-circuited, and output-resistance measurements with the input terminals open-circuited or short-circuited. With reference to Fig. 1–2, the input resistance is expressed for the condition of short-circuited output terminals. This is a series-parallel configuration; it follows from previous discussion that, if

4 · Introduction to Design Principles and Parameters

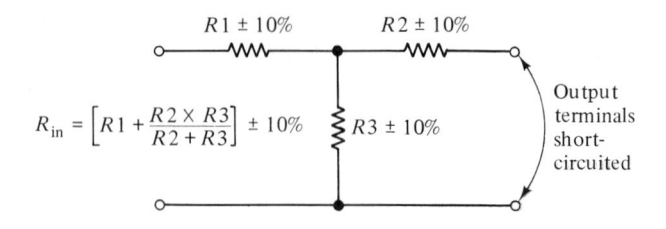

$$R_{in} = \left[R1 + \frac{R2 \times R3}{R2 + R3}\right] \pm 10\%$$

with $R1 \pm 10\%$, $R2 \pm 10\%$, $R3 \pm 10\%$, Output terminals short-circuited.

Figure 1-2 Example of tolerances on a T pad.

the T pad is formed from resistors with the same tolerance rating, the input and output resistance values of the T pad will have the same tolerance as the individual resistors.

1–4 HOMOLOGOUS AND HETEROLOGOUS TOLERANCES

Homologous tolerances comprise the same kind of tolerance on the same kind of component. To continue the foregoing discussions, the resistive tolerances on composition resistors connected in series exemplify homologous tolerances. Heterologous tolerances, on the other hand, comprise different kinds of tolerances on different kinds of components. Refer to Fig. 1-3. Here, three 10,000-ohm ±1 percent resistors are connected in series to serve as a multiplier for a meter movement. This voltmeter arrangement utilizes a movement that is rated for an indication accuracy of ±1 percent of full scale. Note that the multiplier has a resistive (electrical) tolerance of ±1 percent, whereas the movement has an indication accuracy (mechanical) tolerance of ±1 percent. Electrical and mechanical tolerances are different in kind; in other words, they are heterologous tolerances. These heterologous tolerances are additive. In other words, the accuracy rating of this voltmeter arrangement is ±2 percent of full scale.

To clearly understand the additive nature of the multiplier and movement indication tolerances in Fig. 1-3, note that the voltage applied to the terminals of the movement by the multiplier has a tolerance of ±1 percent. Next, the indication accuracy of the movement is established by checking its response to a precise voltage source. It is evident that this indication accuracy, or tolerance, will necessarily add to or subtract from the voltage that is applied to the terminals of the movement by the multiplier. Accordingly, we add the ±1 percent resistive tolerance to the ±1 percent mechanical tolerance to calculate the in-

Basic Device Tolerances · 5

```
       10 K ± 1%  10 K ± 1%  10 K ± 1%   Movement indication
                                         accuracy = ± 1% of
                                         full scale
       ──/\/\/\──/\/\/\──/\/\/\────⊘────┐
       │                                 │
       │                                 │
       │                                 │
       │                                 │
       └──────o Voltmeter accuracy o─────┘
                = ± 2% of full scale
```

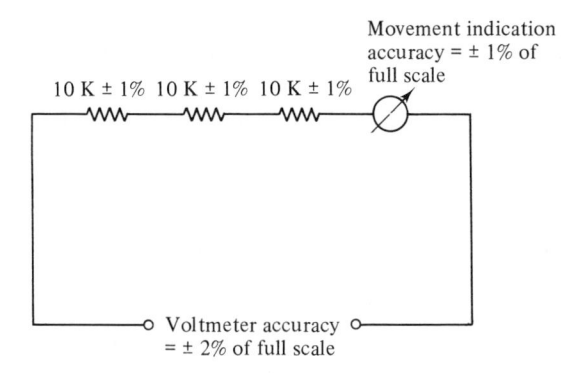

Figure 1–3 Example of voltmeter accuracy rating.

dication accuracy of the voltmeter. This indication accuracy is equal to ±2 percent of full scale. For example, if the full-scale indication of the voltmeter is 100 volts, its corresponding indication accuracy is rated at ±2 volts. Thus, if the scale indication were 50 volts, the value of the applied voltage would fall in the range from 48 to 52 volts.

1–5 BASIC DEVICE TOLERANCES

Device tolerances are seldom rated in terms of a percentage of a bogie value. Occasionally, a certain characteristic will have a specified tolerance. For example, some types of field-effect transistors that are used in direct-coupled applications are tested in production to ensure that the spread of drain current between individual transistors is limited to a variation of approximately two to one. However, it is customary to characterize transistors in terms of maximum ratings on supply voltages, collector and emitter current, power dissipation, maximum operating temperature, and maximum lead soldering temperature. These maximum ratings are generally supplemented by ratings on collector-junction breakdown voltage, maximum collector cutoff current, maximum emitter cutoff current, maximum cutoff frequency, minimum current-gain value, maximum output capacitance, and typical thermal resistance.

In various cases, typical operation in a basic circuit, such as a 455-kHz amplifier configuration, will be provided. The values listed are bogie assignments, and may be stated in terms of maximum and

6 · Introduction to Design Principles and Parameters

minimum values. For example, a certain type of transistor is specified to have a maximum power gain of at least 51 dB at a collector supply voltage of −6 volts and a collector-emitter voltage of −5.7 volts. Further, this same type of transistor is specified to have a maximum gain of at least 54.4 dB at a collector supply voltage of −12 volts and a collector-emitter voltage of −11 volts. These data are of central importance to the circuit designer in calculation of worst-case performance. Forbidden regions of operation for a typical transistor are shown in Fig. 1–4.

Diodes are similarly characterized in terms of maximum ratings on supply voltage, peak reverse voltage, forward current, and lead temperature. The maximum permissible instantaneous voltage drop at the maximum forward-current value is typically specified, with the maximum reverse current that can flow at a stated temperature. The circuit designer should take maximum ratings into consideration so that at no time will any absolute-maximum value be exceeded. This consideration involves the worst probable operating conditions with

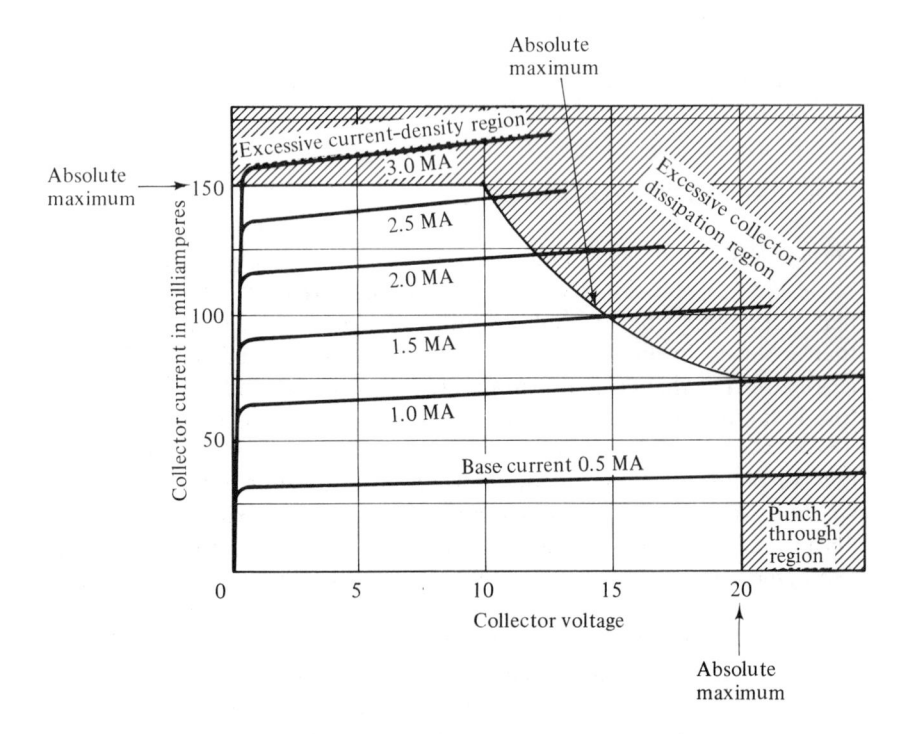

Figure 1–4 Forbidden regions of operation are shown as shaded areas.

respect to supply-voltage variation, equipment component variation, control adjustment, load variation, input signal level changes, unusual environmental conditions, plus all anticipated variations in device characteristics.

Diodes and transistors are available in *matched pairs*. This term denotes that a pair of devices has been selected by the manufacturer for virtually identical characteristics. For example, a matched pair of diodes will have practically the same voltage-current characteristic. On the other hand, another pair of matched diodes from the same production lot will probably have a somewhat different voltage-current characteristic. Matched pairs are used in circuit design to maintain a balanced circuit action. For example, a matched pair of diodes is utilized in a phase-detector circuit; matched pairs of resistors are also often used. As another example, a matched pair of transistors may be required in some complementary-symmetry amplifier designs, to maintain a satisfactorily low distortion level.

1–6 NOTES ON NEGATIVE FEEDBACK

Tolerances on components and devices become less significant when circuits are designed with substantial negative feedback. For example, with reference to Fig. 1-5, an amplifier that develops an intolerable amount of distortion will have greatly improved performance when a suitable value of negative feedback is included. Of course, there is a *trade-off* associated with the use of negative feedback, because the input signal level must be increased to overcome the signal loss owing to negative feedback. In other words, the output level is decreased by negative-feedback action, unless the input signal level is increased to compensate for this loss. Note in passing that there is one kind of device distortion that cannot be reduced by means of negative feedback. Cutoff distortion, caused by driving a transistor beyond its cutoff point, for example, is unaffected by negative-feedback action, inasmuch as the amplifier has zero reserve gain past its cutoff point.

The class of integrated circuits termed *operational amplifiers* ordinarily operate with a very large amount of negative feedback. Basic examples of op-amp amplifier and differentiator arrangements with a large amount of negative feedback are shown in Fig. 1-6. A large amount of negative feedback makes the input-output relation of the configuration virtually independent of op-amp tolerances. As seen in the accompanying equations on the diagrams, the input-out-

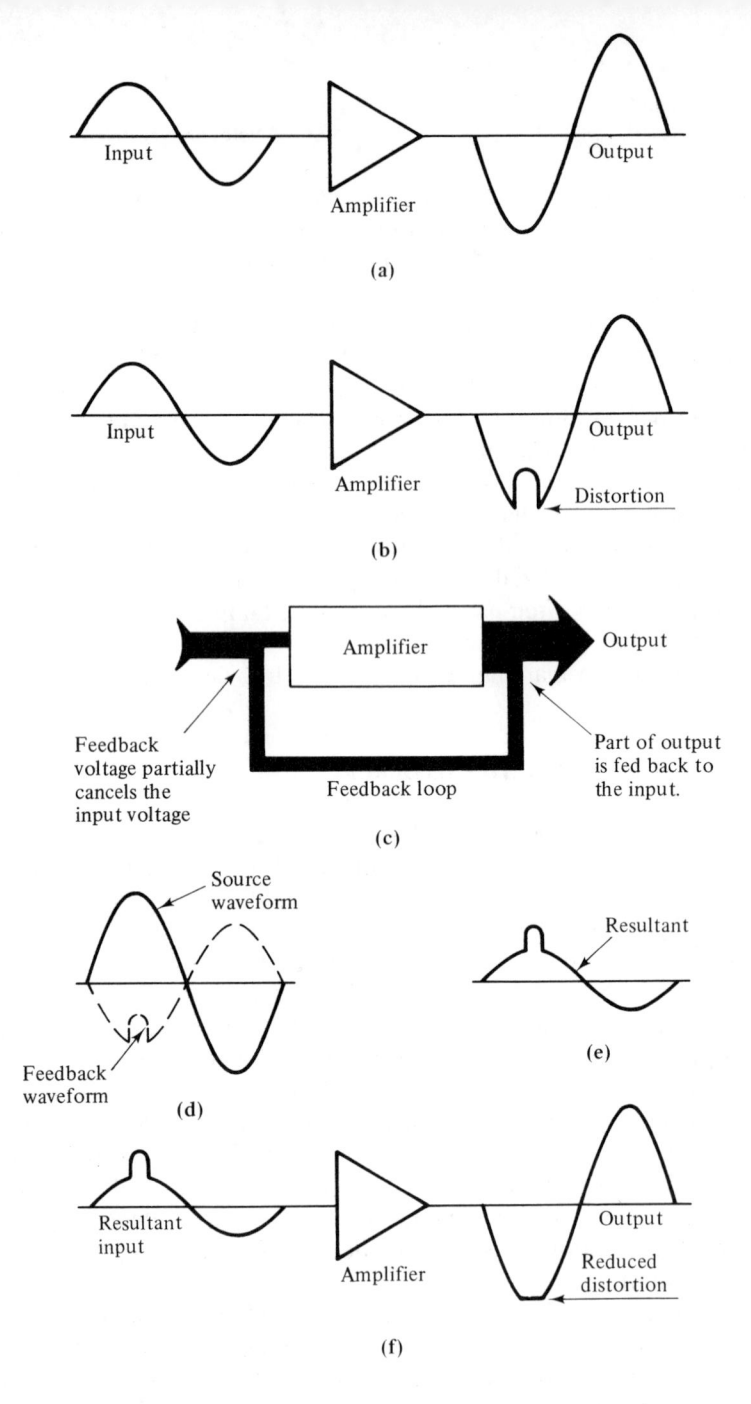

Figure 1-5 Examples of amplifier operation, with and without negative feedback: **(a)** distortionless output; **(b)** distorted output; **(c)** negative-feedback arrangement; **(d)** feedback waveform mixed with source waveform; **(e)** resultant input waveform; **(f)** output waveform has reduced distortion.

8

Amplifier

$$\text{Gain} = \frac{R_0 + R1}{R1}$$

Differentiator

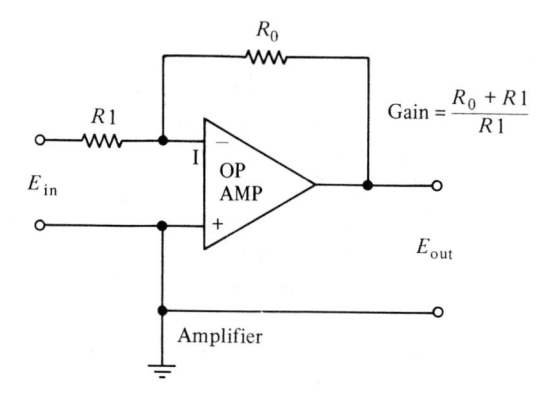

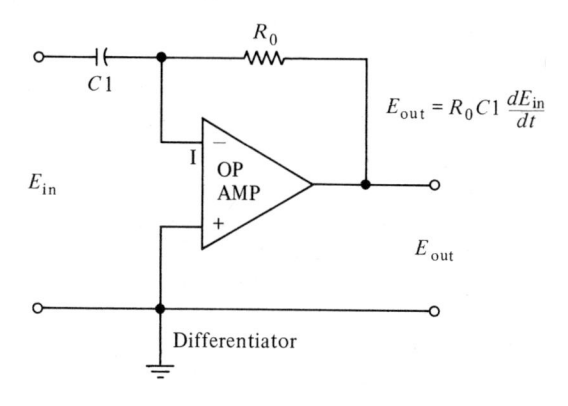

$$E_{out} = R_0 C1 \frac{dE_{in}}{dt}$$

Figure 1-6 Examples of op-amp characteristics when a large amount of negative feedback is used.

put relation becomes practically a function of the component values to which the op amp is connected. In other words, a large amount of negative feedback makes the input terminal *I* "look like" a ground point (virtual ground). Accordingly, the input resistance of the amplifier arrangement is practically equal to R1. The input impedance of the differentiator arrangement is practically equal to the reactance of C1. Of course, if an op amp deteriorates seriously, the amount of negative feedback decreases substantially, and the input terminal *I* no longer approximates a virtual ground.

1-7 POWER DISSIPATION PRINCIPLES

A resistor is rated for maximum power dissipation. A resistor may also be rated for maximum voltage drop. If a 1-megohm resistor has

1-watt construction and a 600-volt maximum rating, it cannot dissipate more than 0.36 watt without exceeding its voltage rating. If resistors are connected in series, the permissible voltage drop of the combination is increased, and the permissible power dissipation of the combination is increased. As an illustration, if two 100,000-ohm, 1-watt resistors are connected in series, the maximum power dissipation becomes 2 watts. Again, if each resistor has a 300-volt maximum rating, the series combination will have a 600-volt maximum capability.

On the other hand, if a 150,000-ohm, 1-watt resistor is connected in series with a 50,000-ohm, 1-watt resistor, the power-dissipation capability of the combination is *not* 2 watts. Since both resistors carry the same current, the power dissipated by the 150,000-ohm resistor is three times the power dissipated by the 50,000-ohm resistor. In other words, the maximum power capability of this series combination is 1.33 watts. The series-connected resistors form a voltage divider, and the voltage drops across the individual resistors are in proportion to their resistance values. Thus, the 150,000-ohm resistor drops 75 percent of the applied voltage, and the 50,000-ohm resistor drops 25 percent of the applied voltage.

When resistors are connected in parallel, the permissible voltage drop for the combination is no different from that of any individual resistor. On the other hand, the permissible power dissipation of the combination is greater than that of any individual resistor. For example, if two 50-ohm, 5-watt resistors are connected in parallel, their 25-ohm combination has a power-dissipation capability of 10 watts. However, if a 30-ohm, 5-watt resistor is connected in parallel with a 150-ohm, 5-watt resistor, this parallel combination does *not* have a power-dissipation capability of 10 watts. Since both resistors have the same voltage drop, the 30-ohm resistor dissipates five times as much power as the 150-ohm resistor; the maximum power capability of this parallel combination is only 6 watts.

A potentiometer is rated for maximum power dissipation, such as 1 watt. This rating applies to the total resistance element. For example, if a 1-watt potentiometer has a total resistance of 2500 ohms, it is permissible to apply 50 volts across the 2500-ohm element. On the other hand, if the potentiometer is set to its 500-ohm position, it would *not* be permissible to apply 50 volts across this smaller resistance value. A maximum of 10 volts could be applied across the 500-ohm portion of the element without exceeding its basic power rating. Next, refer to Fig. 1-7. Here, a potentiometer operates into a comparatively low-resistance load. If the potentiometer is rated for a maximum power dissipation of 1/4 watt, this rating will be exceeded

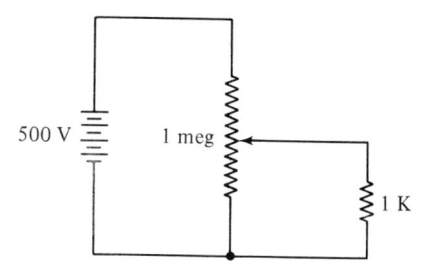

Figure 1-7 Potentiometer circuit with a comparatively low-resistance load.

at any position other than minimum setting. For example, at its midpoint setting, the upper half of the potentiometer's resistance element dissipates practically 1/2 watt. In turn, the circuit designer must specify a much greater power rating for the potentiometer.

1-8 LAW OF NORMAL DISTRIBUTION

Various circuit-design considerations are based on the law of normal distribution. This law defines the range of probabilities for a random occurrence. For example, an extreme worst-case occurrence in a production operation is a random occurrence that has a certain probability. In other words, it is not highly probable that all component and device tolerances will happen to have their worst-case values in a random unit of equipment. Nevertheless, there is this possibility, and it is sometimes possible to calculate the probability of a worst-case occurrence in production. The circuit designer may decide to specify component and device tolerances which ensure that an occasional worst-case occurrence will still be within acceptable performance limits. Again, the designer may decide to relax component and device tolerances, and to take a calculated risk that the number of production rejects will not increase overall manufacturing costs objectionably.

A standard probability curve is shown in Fig. 1-8. In this example, the curve indicates in a qualitative way the probability of a worst-case occurrence versus the number of production units. In other words, if an unlimited (theoretically infinite) number of units are produced, there is a 100 percent probability of a worst-case occurrence. Conversely, if very few units are produced, the probability of a worst-case occurrence is very small. Worst-case occurrences in

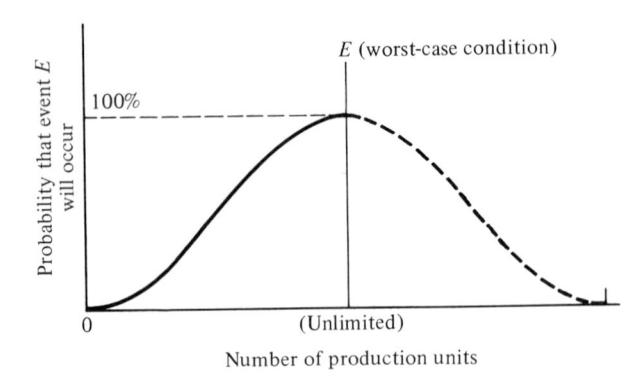

12 · Introduction to Design Principles and Parameters

Figure 1-8 A qualitative example of the probability of a worst-case occurrence.

a production operation can be treated in terms of standard probability calculations only when component and device tolerances are truly random. It oftens happens that tolerances in a parts shipment tend to cluster toward the low end or the high end of the specified limits. In ordinary commercial design projects, the engineer must simply assume that component and device tolerances are truly random.

1-9 NOTES ON MATHEMATICAL MODELS

Electrical and electronic signs and symbols have meanings that must be properly understood if they are to be used effectively. Signs and symbols may have either one of two basic classes of meaning. A sign or a symbol may represent a physical property, such as voltage, current, or resistance. On the other hand, a sign or symbol may represent a mathematical concept, such as an independent or a dependent variable. Laws of electricity and electronics express observed (experimental) relations among physical concepts that are quantitatively related in satisfactory accordance with an algebraic equation. For example, Ohm's law is expressed by a simple algebraic equation.

It is evident, even to the most inexperienced circuit designer, that the physical significance attributed to any equation is subject to various limitations. These limitations may be comparatively few in some cases, and they may be very extensive in other cases. In turn, a mathematical model must be regarded as a potentially treacherous, if powerful, design tool. Only the physical meaning of a symbol is considered when a law of electricity or electronics is stated in general terms. On the other hand, to solve a circuit problem the designer considers the mathematical meaning of the symbol within the context of the equation in which it appears. He performs mathematical operations on basic equations, relates the equations, expresses them in various forms, and explicitly derives a relation between variables that was only implicit in the basic equations.

After the circuit designer has obtained a mathematical solution to his problem, he ceases to regard a symbol as a mathematical variable, and considers it to be a quantitative physical entity. *Because the rules for mathematical operations are not necessarily the same as the laws that relate modeled physical concepts, there is always some hazard that the circuit designer will obtain nonphysical, spurious, nondescriptive, or conditionally valid answers to the problem under consideration.* Therefore, the designer must always be on guard for nonphysical answers, and reject them out of hand as soon as they are detected. The circuit arrangement depicted in Fig. 1–9 illustrates a basic distinction between a physical circuit and a mathematical model of the circuit—if a current of two amperes flows in the wire, it is obvious that the number is not in the wire. Neither is the number on the scale of the ammeter—there is a numeral "2" that appears on the ammeter scale, and this numeral (a physical entity) corresponds to the number 2 (a mathematical concept in the circuit-designer's mind).

A simple example of an ordinarily (though not necessarily) nonphysical answer is encountered in solving for R_0 in the configura-

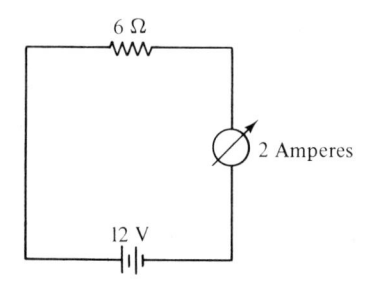

Figure 1–9 If a current of two amperes flows in a wire, the number is not in the wire.

14 · Introduction to Design Principles and Parameters

tion of Fig. 1–10. This is a resistive series-parallel circuit, and its input resistance is expressed

$$R_0 = R_1 + \frac{R_2 R_0}{R_2 + R_0}$$

To solve for R_0, the foregoing equation is written in standard quadratic form:

$$R_0^2 - R_1 R_0 - R_1 R_2 = 0$$

R_0 is then expressed by the relation:

$$R_0 = \frac{R_1 \pm \sqrt{R_1^2 + 4R_1 R_2}}{2}$$

In turn, two answers are obtained; one answer denotes a positive resistance, and the other answer denotes a negative resistance:

$$R_0 = \frac{R_1 + \sqrt{R_1^2 + 4R_1 R_2}}{2} \qquad \text{(positive answer)}$$

and $\quad R_0 = \dfrac{R_1 - \sqrt{R_1^2 + 4R_1 R_2}}{2} \qquad$ (negative answer)

Ordinarily, the circuit designer will be working with positive resistance values, and the negative answer will be rejected as nonphysical. However, if the designer were working with tunnel diodes, for example, he could regard the negative answer as the solution to a certain kind of circuit problem. The physical behavior of a circuit is *modeled* mathematically by equations that have physical interpretations, such as Ohm's law. This relation between a circuit's behavior and any descriptive equation is one of *analogy*. Physical analogies are also invoked in circuit design. For example, electric current (charge flow) in wires is sometimes regarded as analogous to water flow in pipes. Voltage is sometimes regarded as analogous to water

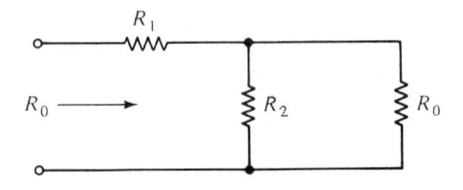

Figure 1–10 A positive answer and a negative answer are obtained when one is solving for the value of R_0.

pressure. However, such analogies lead to absurd conclusions if pursued too far. This pitfall is termed "exceeding the competence of the model."

1–10 COMPUTER-AIDED DESIGN

Computer-aided design denotes the solution of comparatively involved mathematic problems by means of a computer. Either analog or digital computers may be used, although most circuit-design problems are processed by digital computers. Either a calculator or a large computer may be utilized, depending on the complexity of the problem. If a circuit design problem cannot be solved readily on a slide rule, but is not unduly complex, it can often be satisfactorily processed by a pocket calculator, such as that illustrated in Fig. 1–11. If the operations required in a mathematical problem cannot be readily performed with a calculator, a large digital computer would be employed.

Figure 1–11 A pocket calculator is capable of solving many engineering design problems. *(Courtesy, Hewlett-Packard)*

1–11 PITFALLS IN DESIGN PROCEDURES

Various pitfalls await the inexperienced or the unwary circuit designer. The danger of an uncritical extension of a circuit analogy has been previously noted. Even a prototype model will often conceal seductive pitfalls. For example, if the prototype performs as planned, and if it has been assembled from near-bogie component and device values, it is tempting to relax vigilance and to assume that the prototype will be completely representative of the final mass-produced equipment. However, the critical designer will proceed to check all operating characteristics of the prototype model. In other words, he will assume that some physical factor is likely to have been overlooked, or that some characteristic has been miscalculated. As an illustration, all electronic equipment generates heat; however, the internal operating temperature rise above ambient is difficult to calculate accurately. Accordingly, a careful designer proceeds to check equipment operation in the highest specified ambient temperature, and to measure the temperature at each component and device that is rated for a maximum operating temperature.

Another pitfall that awaits the unwary circuit designer is neglect of the power-supply characteristics. For example, an amplifier that operates satisfactorily from a bench power supply may or may not remain stable when energized from a common power supply that also energizes other amplifiers in a system. In other words, prototype operation should be checked under system conditions. It is possible that the designer will encounter system instability owing to common power-supply impedance. In this situation, adequate decoupling would have to be included in the prototype model for rigorous power-supply isolation.

As another example of a pitfall in the path of the circuit designer, there is sometimes a temptation to make use of nonspecified but attractive device characteristics. In turn, simplified design could backfire as 100 percent rejection of future production runs. Thus, digital-logic designers have two varieties of *JK* flip-flops available. With reference to Fig. 1–12, one is termed a *negative edge-triggered JK flip-flop,* and the other is called a *ones-catching flip-flop.* Edge-triggering denotes that the flip-flop output is a function of the input states only at the time of the clock transition. Note that a ones-catching flip-flop operates satisfactorily as a *JK* flip-flop and is generally described as edge-triggered in data sheets. Nevertheless, a

Pitfalls in Design Procedures · 17

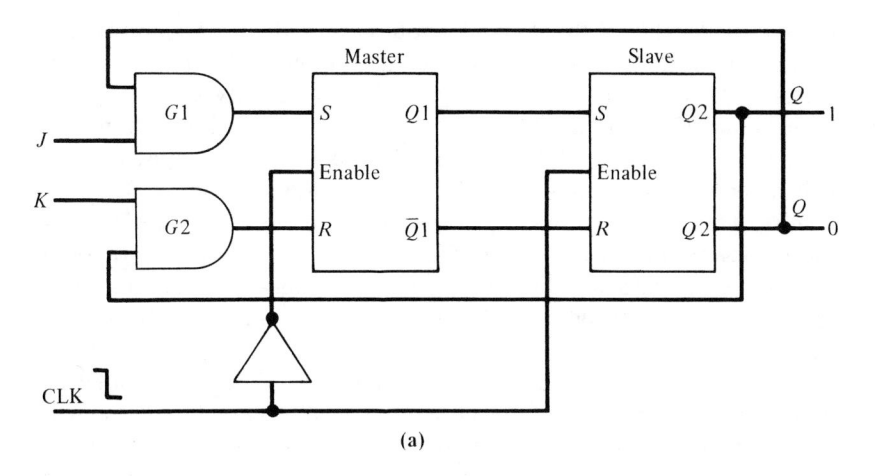

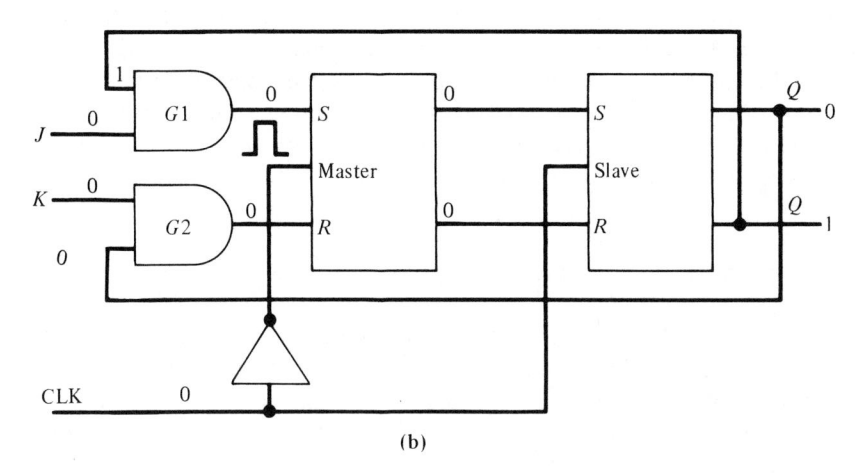

Figure 1–12 Two types of flip-flops: **(a)** negative edge-triggered *JK* flip-flop; **(b)** ones-catching flip-flop. *(Courtesy of Hewlett-Packard)*

ones-catching flip-flop is not edge-triggered in the strict sense of the term. This distinction is detailed as follows.

Consider the response of a negative edge-triggered JK flip-flop. During the time that the clock is low, the master is enabled and its outputs Q1 and $\overline{Q}1$ assume the *J* and *K* input states. Next, during the time that the clock is high, the master is frozen at its last input state. Meanwhile, the slave is enabled and it assumes the states of Q1 and $\overline{Q}1$. The feedback loops from slave to master ensure that the flip-flop will toggle if the J and K inputs are both 1. Compare this response with that of the ones-catching flip-flop. It operates satisfactorily as a JK flip-flop, although it is not actually edge-triggered. If the J and K inputs are both zero, and the clock is low, the master assumes a 00 state. Note that if the $\overline{Q}$ output is 1, the input gate G1 will respond to a pulse on the J line and a momentary 10 SR input results. This input condition leaves the master in a 10 state, and the slave toggles incorrectly when the clock pulse is applied. In other words, a ones-catching flip-flop is not truly edge-triggered, inasmuch as its output is a function of the JK inputs before and during the edge transition.

If the digital circuit designer chooses to simplify a configuration by pulsing the J line of a ones-catching flip-flop while the master is in a 00 state, he utilizes a nonspecified characteristic of the flip-flop. This is a precarious procedure, because even if one manufacturer has a policy of retaining the ones-catching characteristic of a particular "edge-triggered" flip-flop, the policy of a second source will be unknown. In turn, the circuit designer places himself in a straitjacket from the viewpoint of *device procurement*. The calculated risk that is involved would be declined by the vast majority of experienced circuit designers.

1–12 HIDDEN VARIABLES

Hidden variables are sometimes encountered in the course of circuit-design procedures. For example, narrow pulses, or similar short-duration "glitches" may cause faulty operation in digital circuits. A typical glitch waveform is shown in Fig. 1–13. The source of a glitch may be radiated noise, power-line coupled noise, sliver pulses resulting from timing-skew input signals to gates, or other marginal circuit operations. To capture these rapid transients on an oscilloscope screen is sometimes a very difficult task, even if the circuit designer knows that they are present. Again, if it is only suspected that glitches may be the cause of malfunction, the designer's difficulties are increased. When glitches occur at a low repetition rate, even most of the conventional fast-writing oscilloscopes can display only a very dim pattern. If the circuit designer is confronted by this problem, he should utilize a storage-type oscilloscope with a fast writing rate. A fast-storage function builds up a barely visible, or an invisible, display to a clearly viewable level.

Another hidden variable that can be troublesome on occasion is

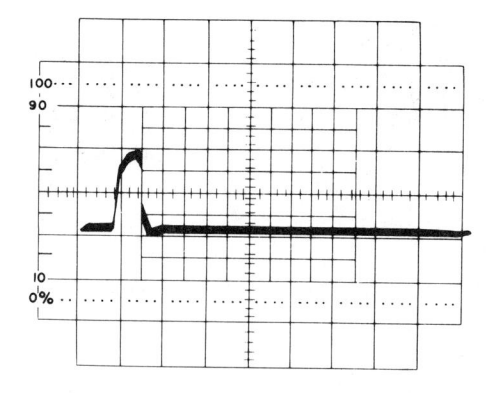

Figure 1-13 A 25-nsec glitch. *(Courtesy of Tektronix)*

AC current flow through an electrolytic capacitor that is operated near its rated value of DC voltage. If the capacitor were a pure reactance, AC current flow would dissipate no power, and would not cause the temperature of the capacitor to rise. However, an electrolytic capacitor is an impedance, and its power factor tends to increase as its temperature increases. If the frequency of the AC current is comparatively high with respect to the impedance of the capacitor, a large AC current flow will develop a relatively small voltage drop across the capacitor. In other words, the circuit designer must apply a suitable derating factor to an electrolytic capacitor that is operated near its rated DC working voltage and temperature limit, in the event that it is also required to carry a substantial value of AC current.

Still another type of hidden variable in some circuit-design projects concerns possible adverse environmental conditions in which the equipment may be operated. For example, a radio receiver is ordinarily operated inside of a residential building, where it is protected from dew, fog, rain, salt spray, and excessive shock or vibration. However, it is possible that some set-owners will operate the receiver on a patio, where it is exposed to nightly dew or fog. Other set-owners may operate the receiver in a motorboat, where it is subjected to occasional or repeated salt-water spray or splashes. Still other set-owners may operate the receiver in a commercial vehicle under conditions of substantial shock and vibration. This hidden variable can be difficult to assess in view of increased production costs, and the circuit designer is advised to consult with the

20 · Introduction to Design Principles and Parameters

Figure 1-14 Typical Underwriters' Laboratories labels.

sales department regarding anticipated markets, and to seek an executive-level decision for establishment of production-cost guidelines.

1-13 UNDERWRITERS' LABORATORIES REQUIREMENTS

Various specifications have been established by the Underwriters' Laboratories (UL) for the reduction of shock and fire hazards in the design of electrical and electronic equipment. The Underwriters' Laboratories is a branch of the National Board of Fire Underwriters. A consumer electronics product that has been examined by the UL, and which has been accepted, is permitted to carry a tag or label of approval, such as that shown in Fig. 1-14. Circuit designers can obtain a pamphlet from any UL office to determine the most recent electrical and mechanical specifications that are required for approval by UL inspectors. Specific approval of each design and model is necessary. Many municipalities require UL approval for all products sold in their jurisdictional area.

It is instructive to note some basic UL requirements. For example, a power transformer used in a consumer-electronics product must be completely enclosed in a metal or other noncombustible housing to prevent the escape of flames or molten metal in the event that the transformer were destroyed by a catastrophic overload. A power transformer with a UL-approved housing is depicted in Fig. 1-15. As another illustration, if an AC/DC power supply is utilized, with one side of the power line connected to a metal chassis ("hot-chassis") design, a substantial back cover is required with an interlock. A polarized type of power plug is also stipulated. Adequate operator protection must be provided against the possibility of shock from exposed metal parts such as ornamentation, setscrews, shafts of control knobs, assembly bolts or screws, cabinet, antenna and/or ground lead(s), or the tone arm in a record player.

Figure 1-15 A power transformer with a UL-approved housing.

A circuit breaker or fuse must be included in the input line of the power supply to minimize the danger of fire in the event of a serious overcurrent demand anywhere in the equipment circuitry. Temperature limits for various classes of materials must be observed. Each conductor is required to have adequate cross section and insulation for its intended current flow, operating temperature, and application. Power cords must meet specified requirements for stranded conductors at the rated current demand, and must be adequately insulated. A power cord must be provided with strain relief inside of the equipment; the cord must enter through an insulating bushing that will prevent cutting or fraying of the insulation on the cord. Also, the power cord must be secured at its entry point so that it cannot be pushed back into the interior of the cabinet. Equipment that operates at 25,000 volts or more must include adequate shielding for protection against X-ray hazards, with fail-safe facilities.

1-14 ENVIRONMENTAL TESTING

Environmental testing is defined as the testing of a system or component under controlled environmental conditions, each of which tends to affect its operation or its life. Environmental conditions are external conditions such as heat, shock, vibration, pressure, moisture, dust, radiation, lightning, and so on. As an illustration, a high-voltage circuit that operates satisfactorily at sea level may break down immediately when installed in a bomber that flies at high altitudes. The circuit designer will make environmental tests in this example with the aid of an environmental chamber in which the air pressure can be reduced as required. Water vapor may be admitted into the chamber to check circuit operation under conditions of high humidity. The atmosphere inside the chamber may be cooled to a very low temperature, or heated to a very high temperature. The capability of a design to withstand shock and vibration is often checked by securing the unit to an oscillating shock table, and subjecting it to a chosen number of *g*'s at a selected frequency for a specified period of time. (A *g*, or g-force, denotes the acceleration due to gravity, or 32.17 ft/sec^2.)

1–15 BURN-IN DESIGN TEST PROCEDURES

Burn-in denotes the operation of a device to stabilize its failure rate. This term also applies to a circuit board, for example. This test procedure provides design verification during the *early-failure period*, starting immediately after assembly, during which equipment failures initially occur at a higher than normal rate owing to the presence of defective devices or components, to improper production processes, or to inadequate design. This early-failure period is also called the *debugging period,* or *infant-mortality* period. A burn-in period involves continuous operation of a unit at maximum rated power level, for a length of time that has been established by experience. In other words, when the mortality rate is plotted against time, this rate typically decreases and approaches zero as a limit. Design is judged adequate, and production processes considered appropriate, when the initial burn-in failure rate is very small.

1–16 LIFE TESTS

Life testing denotes a determination of the normal *life expectancy* of a device, component, or circuit arrangement during normal use. Tests are made under conditions that approximate, or that simulate by acceleration, a lifetime of normal use. *Accelerated life tests* typically involve operations such as repeatedly turning the power switch of a unit on and off several times a minute, for days, weeks, or months, until a failure occurs. Burn-in tests are also a form of life test insofar as the infant-mortality period is concerned. A life test may be made inside of an environmental chamber. For example, a high-voltage circuit is sometimes short-lived when cycled repeatedly at low temperature and low pressure, alternated by cycling at normal temperature and air pressure. Life tests serve to reveal design deficiencies, and also provide data for *reliability* evaluation. This evaluation is expressed as a probability conclusion that the design will perform adequately for the length of time intended and under the operating environment that is encountered.

Reliability control denotes the coordination and direction of technical reliability activities from a system viewpoint. Data related to the frequency of failure of a component, device, circuit, or system are termed *reliability data*. These data may be expressed in terms of *failure rate,* mean time between failure *(mtbf),* or *probability of success.* In a complex circuit system, the data may be contained in comprehensive documents that provide a detailed history of the reliability evaluation of parts, assemblies, etc., or the entire program during the design, development, production, and major product improvement phases of an equipment in which engineering studies have been made to select the most reliable product for the intended application.

A *confidence evaluation* expresses a degree of assurance that a stated failure rate has not been exceeded. It is given as a percentage, called the *confidence factor,* that denotes the evaluated degree of assurance. The *confidence interval* states a range of values that are believed to include, with a preassigned degree of confidence, the true characteristic of a production lot. A *confidence level* expresses the probability, as a percentage, that a given conclusion is true, or that it lies within certain limits that have been calculated at a certain date. Thus, a confidence level can be described as a degree of certainty. The *confidence limits* are the extremes of a confidence interval within which there is a designated chance that the true value is included.

Note that a *prototype model* is a working model, usually hand-assembled, and suitable for complete evaluation of mechanical and electrical form, design, and performance. Approved parts are employed throughout, so that the model will be completely representative of the final, mass-produced equipment. The *producer's liability risk* is the risk faced by the producer (usually set at 10 percent), that a product will be rejected by a reliability-acceptance test, even though the product is actually equal to or better than a specified value of reliability. *Production sampling tests* are the tests normally made by either the vendor or the purchaser on a portion of a production lot, for the purpose of determining the general performance level. *Production tests* are those tests normally made on 100 percent of the items in a production lot by the vendor, and ordinarily on a sampling basis by the purchaser.

Quality denotes the extent of conformance to specifications of a device or the proportion of satisfactory devices in a lot. *Quality assurance* is the planned, systematic pattern of actions necessary to provide suitable confidence that an item will perform satisfactorily in actual operation. *Quality control* consists of means to control the variation of workmanship, processes, and materials in order to produce a consistent, uniform product. *Quality engineering* denotes an engineering program, the purposes of which are to establish suitable quality tests and quality acceptance criteria and to interpret quality data.

Consider a basic example of a reliability level rating for a resistor. This rating is expressed as a percentage per 1000 hours. In other words, the manufacturer has made extensive life tests of this type of resistor, and has established that not more than the rated percentage of a production lot will be found out of tolerance after 1000 hours of normal operation according to strict statistical analysis. This reliability level rating is indicated by a fifth colored band on a resistor. Significance of various colors is as follows:

Brown, 1 percent reliability level
Red, 0.1 percent reliability level
Orange, 0.01 percent reliability level
Yellow, 0.001 percent reliability level

Again, consider a typical life test for a metallized polypropylene fixed capacitor. A life test is made for 1000 hours at 105°C and at 150 percent of rated voltage. It is stipulated that the insulation resistance and the dissipation factor shall remain within initial tolerance ratings, and that not more than one failure will occur for each 12 capacitors undergoing the life test. When the capacitor is tested under conditions of high humidity, it is stipulated that the insulation resistance shall not decrease more than 20 percent below rated minimum value, that the dissipation factor shall not exceed 0.1 percent, and that the capacitance value shall not change more than 0.5 percent. It is stipulated that not more than one failure will occur for each 12 capacitors undergoing the high-humidity test.

An example of 95 percent confidence limits for noise as a function of resistance value for carbon film resistors is shown in Fig. 1–16. Thus, there is 95 percent chance that a 1-megohm, 1/4-watt resistor of this manufacture will have a noise level less than 3 μV/V. Next, an example of a 90 percent confidence interval for the temperature coefficient of a carbon film resistor as a function of the resistance value is shown in Fig. 1–17. In other words, a 1-megohm resistor of this manufacture is rated for a temperature coefficient in the range from −299 to

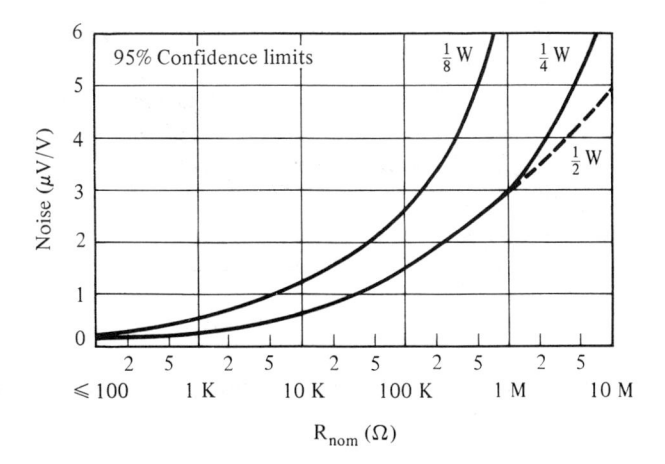

Figure 1-16 Noise as a function of the resistance value for Ohmite "Little Rebel" carbon film resistors. *(Courtesy, Ohmite Mfg. Co.)*

−450 ppm/°C with a 90 percent degree of confidence from the viewpoint of the theory of probability.

A carbon film resistor has some response to temperature cycling, although the response is normally small. For example, a typical resistor of this type is rated for a resistance change of less than 0.5 percent or 0.5 ohm, when cycled for five times at a temperature of −55°C for three hours, followed by a temperature of +155°C for three hours. Again, after long-term exposure to damp heat at 40°C, this type of resistor is rated for a resistance change of less than 3 percent. For example, after exposure for 56 days to a humidity of 90–95 percent, at a potential difference of 5 volts DC, a resistor with a nominal value in the range from 10 ohms to 100 kilohms is rated for a resistance change of not more than 1 percent. In the range from 100 kilohms to 1 megohm, it is rated for a resistance change of not more than 2 percent. In the range above 1 megohm, it is rated for a resistance change of not more than 3 percent.

The circuit designer should also take lead-test ratings into consideration. As an illustration, tensile, bending, and torsion stresses and strains on the resistor leads, short of physical damage, are warranted to cause a resistance change no greater than 0.5 percent or 0.5 ohm. Shock and vibration tests are related to lead tests. Thus, a typical resistor is rated for a resistance change of not more than 0.5 percent or 0.5 ohm when subjected to 3 × 1500 50-g shocks in three directions. Similarly, this type of resistor is rated for a resistance change of not more

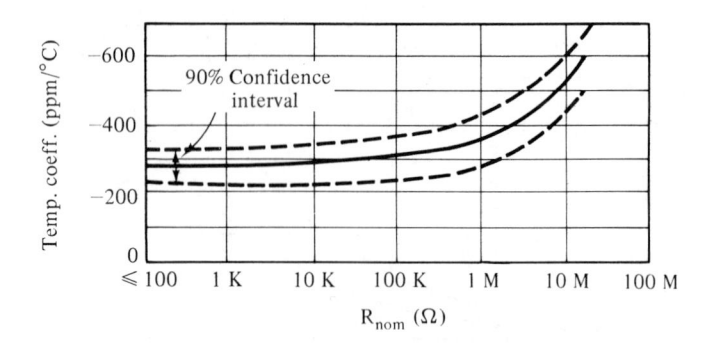

Figure 1-17 Temperature coefficient as a function of the resistance value for Ohmite "Little Rebel" carbon film resistors. *(Courtesy, Ohmite Mfg. Co.)*

than 0.5 percent or 0.5 ohm when subjected to a displacement of 0.06 inch, or an acceleration of 50 g, in three directions for a period of nine hours. Good design practice regards resistance changes from various causes as independent and cumulative.

CHAPTER 2

Elements of Electronic Circuit Design

2–1 RC FILTER PRINCIPLES

A *filter* is defined as a frequency-selective configuration of resistors, capacitors, and/or inductors, with or without active devices, that functions to attenuate or block current flow at some frequencies, with little opposition to current flow at other frequencies. This definition includes zero frequency (direct current). A *passive filter* contains only passive components: resistance, capacitance, and/or inductance. On the other hand, an *active filter* includes an active device, such as a transistor or an integrated circuit. Active and passive filters are also called *wave filters*. RC filters can be subdivided into high-pass, low-pass, bandpass, and band-elimination types. A *low-pass filter* is also termed a *high-cut filter;* a *high-pass filter* is also called a *low-cut filter*. These types are widely used in audio tone controls, as exemplified in Fig. 2–1.

Note that an *RC* high-pass filter section is also used extensively as a *coupling circuit*. In this application, it functions to filter out (block) the DC component of a pulsating-DC waveform, and to pass the AC component of the waveform with negligible frequency discrimination over a specified range. For example, an audio mixer configuration with *RC* coupling circuits is shown in Fig. 2–2. Again, an *RC* low-pass filter section is often used as a *decoupling circuit*. In this application, it functions to greatly attenuate (bypass) the AC component of a waveform, and to pass the DC component of the

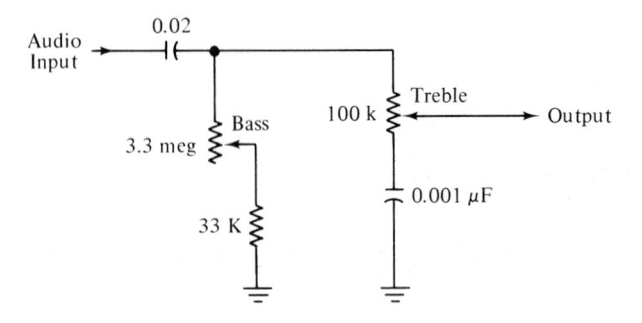

Figure 2–1 A basic audio tone control configuration.

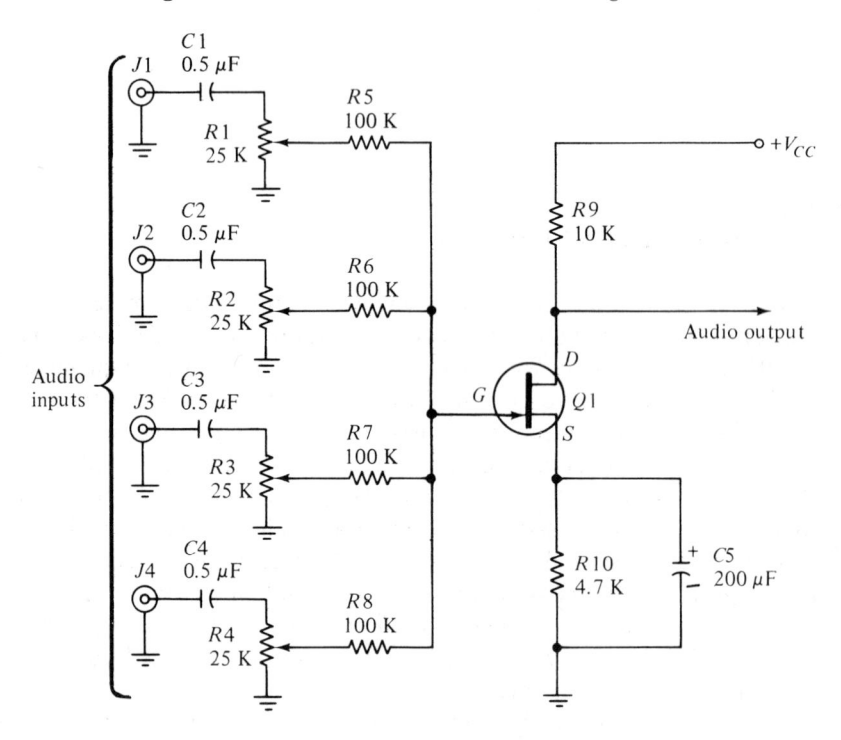

Figure 2–2 Audio mixer with *RC* coupling circuits.

waveform. Low- and high-pass filter action is inherent in any basic *RC* circuit, although low-pass action in one circuit can be *compensated* by high-pass action in another circuit. These basic filters are defined as follows:

> A *low-pass filter* functions to permit current flow at all frequencies below a specified *cutoff frequency* with little (theoretically zero) loss, but to discriminate extensively (theo-

retically completely) against current flow at frequencies above the cutoff frequency.

A *high-pass filter* functions to permit current flow at all frequencies above a specified *cutoff frequency* with little (theoretically zero) loss, but to discriminate extensively (theoretically completely) against current flow at frequencies below the cutoff frequency.

Unless otherwise specified, the cutoff frequency of a filter is defined as the frequency at the −3-dB point on its frequency-response curve. The output voltage from a filter is 70.7 percent of maximum at its −3-dB point, as seen in Fig. 2–3 (page 30).

Component Tolerances In most applications, the filter cutoff frequency is of chief concern. Component tolerances of ±20 percent affect the cutoff frequency of an *RC* low-pass filter as exemplified in Fig. 2–4. The cutoff frequency is equal to $1/(2\pi RC)$. Thus, if $R = 10,000$ ohms, and $C = 0.001$ μF, the cutoff frequency of the filter is 15,920 Hz. This is the design-center or bogie value. If C has its bogie value, and R has a tolerance of ±20 percent, the cutoff frequency ranges from 13,267 Hz to 19,900 Hz. Again, if R has its bogie value, and C has a tolerance of ±20 percent, the cutoff frequency ranges from 13,267 Hz to 19,900 Hz (as before).

However, the circuit designer is most concerned with the *worst-case* condition. In this type of circuit, the worst-case condition occurs when both of the tolerances have their negative limit values. Thus, the worse-case cutoff frequency is 24,875 Hz, compared with a 15,920-Hz bogie value. If both of the tolerances have their positive limit values, the cutoff frequency is 11,055 Hz. This cutoff frequency range may be expressed as 15,920 Hz + 8855 Hz − 4865 Hz. If the worst-case condition is unacceptable, tighter-tolerance components must be utilized.

It is instructive to note that if the tolerance on R is at its positive limit, and the tolerance on C is at its negative limit, the tolerances tend to cancel, although they do not cancel completely. To continue the foregoing example, the cutoff frequency becomes 16,583 Hz, which is 668 Hz greater than the bogie value of 15,920 Hz. Consider the percentage of cutoff-frequency errors that result from 20-percent component tolerances:

1. If C has its bogie value and the value of R is 20 percent high, the cutoff-frequency error is approximately 17 percent. Or, if R has its bogie value and the value of C is 20 percent high, the cutoff-frequency error is approximately 17 percent.

30 · Elements of Electronic Circuit Design

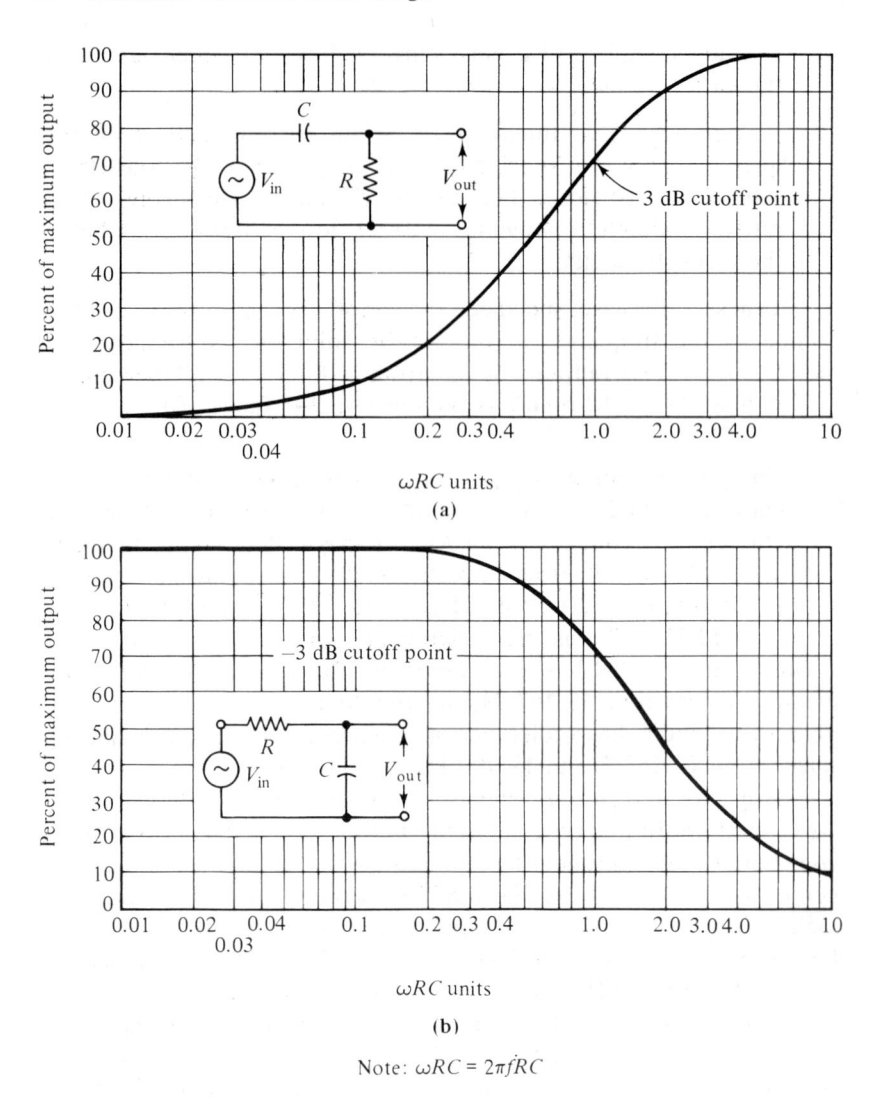

Figure 2-3 Universal frequency response charts for *RC* filter sections: **(a)** high-pass section; **(b)** low-pass section.

Note: $\omega RC = 2\pi f RC$

2. If *C* has its bogie value and the value of *R* is 20 percent low, the cutoff-frequency error is 25 percent. Or, if *R* has its bogie value and the value of *C* is 20 percent high, the cutoff frequency error is 25 percent.
3. If the value of *R* is 20 percent high, and the value of *C* is

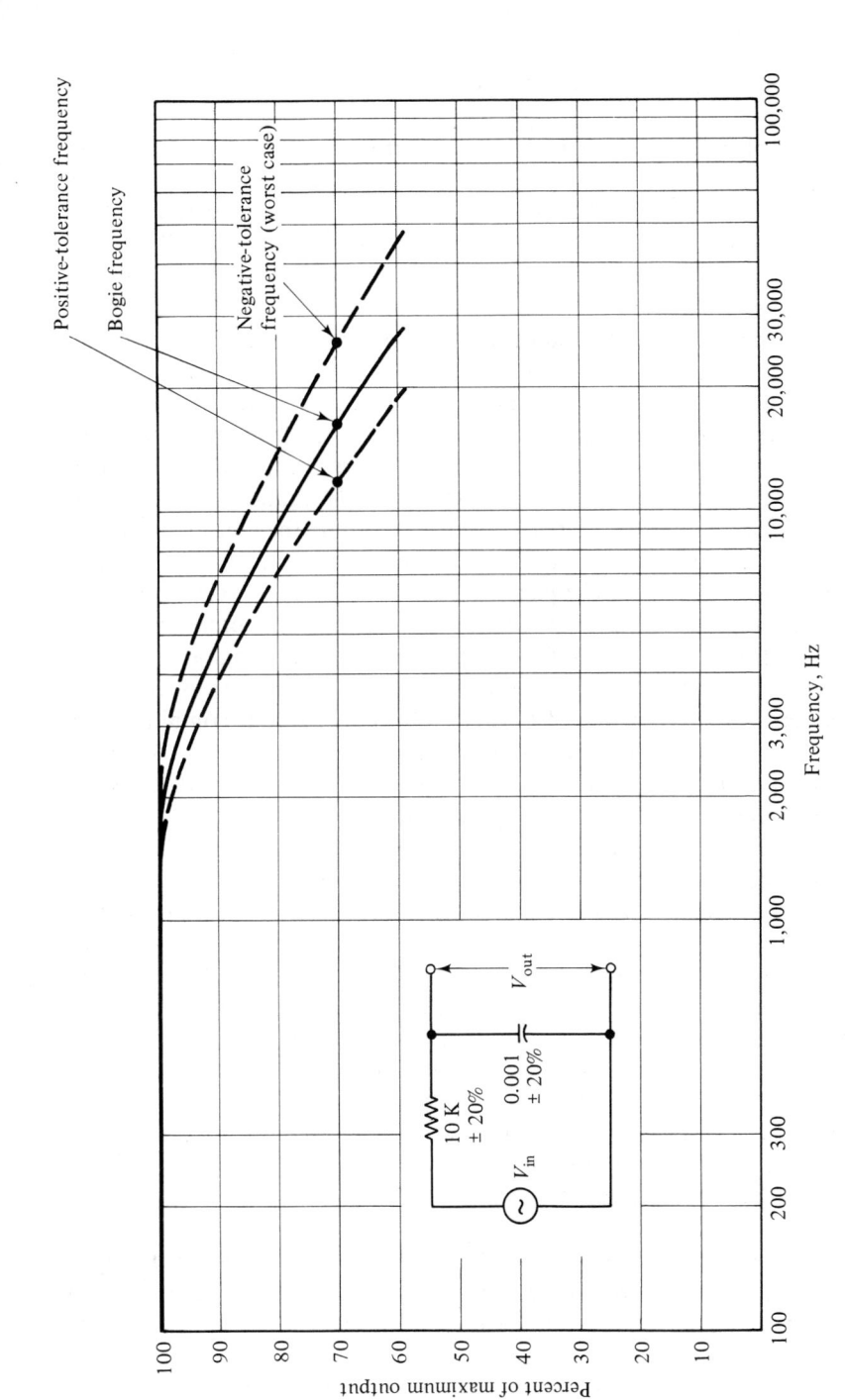

Figure 2-4 Example of ±20 percent component tolerances on cutoff frequency.

also 20 percent high, the cutoff-frequency error is 31 percent.
4. If the value of R is 20 percent low, and the value of C is also 20 percent low, the cutoff-frequency error is 56 percent (worst case).
5. If the value of R is 20 percent high, and the value of C is 20 percent low, the cutoff-frequency error is 4 percent.

Input Impedance The input impedance of either the high-pass or the low-pass filter section depicted in Fig. 2-3 depends upon the value of the load that may be connected across the output terminals of the section. This topic considers the simple situation in which the load value is so high that it can be neglected; the output terminals are in effect open-circuited. In turn, the input impedance of either the high-pass or the low-pass section is infinite at zero frequency (DC), and is equal to R at infinite frequency. To continue the example of Fig. 2-4, the input impedance is 14,142 ohms at the cutoff frequency of 15,920 Hz, as shown in Fig. 2-5. At the −3-dB cutoff frequency, the reactance of C is $j10,000$ ohms, which combines with the 10,000-ohm resistance to form an impedance of 14,142 ohms with a phase angle of 45 deg. Note that the input current leads the input voltage in either the low-pass or the high-pass section.

Consider the effect of component *tolerances* on the input-resistance value. If the resistor has an actual value of 12,000 ohms, and the capacitor has an actual value of 0.0012 μF, the input impedance of the RC filter section at its cutoff frequency becomes 17,000 ohms, as compared with a bogie value of 14,142 ohms. This is an error of 20 percent. Next, if the resistor has an actual value of 8000 ohms, and the capacitor has an actual value of 0.0008 μF, the input impedance of the RC filter section becomes 11,300 ohms, as compared with a bogie value of 14,142 ohms. This is an error of 28 percent, and is the *worst-case value* when the components have tolerances of ±20 percent. In summary, the input impedance of the RC filter section could range from 11,300 ohms to 17,000 ohms.

Output Impedance Consider the output impedance of the low-pass and high-pass RC filter sections depicted in Fig. 2-3. This is the impedance "looking backward" from the output terminals into the configuration. In these examples, a constant-voltage source is assumed; the source has an internal resistance of zero. In turn, the source is considered as a short circuit in an output-impedance analysis. Thus, the output impedance of either section consists of a paral-

RC Filter Principles · 33

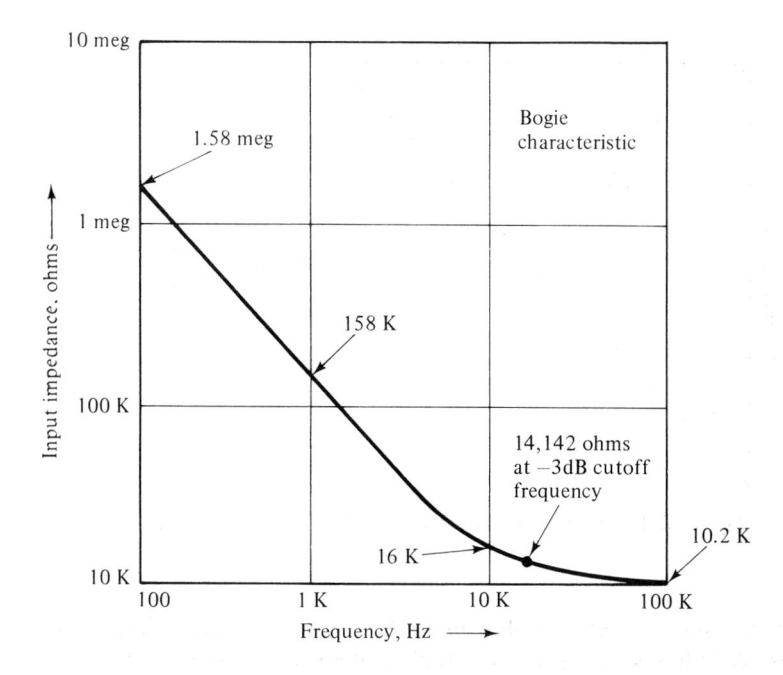

[Graph: Input impedance (ohms) vs Frequency (Hz), showing values 1.58 meg, 158 K, 16 K, 14,142 ohms at −3dB cutoff frequency, 10.2 K. Labeled "Bogie characteristic".]

Figure 2-5 Input-impedance variation of an *RC* high-pass filter section versus frequency.

lel combination of a resistor and a capacitor. To continue with the example of Fig. 2-4, the section has an output impedance of 10,000 ohms at zero frequency (DC), and zero output impedance at infinite frequency. At the cutoff frequency of 15,920 Hz, the 10-kΩ resistor operates in parallel with a 10-kΩ capacitive reactance. In turn, the output impedance at the cutoff frequency is 7072 ohms, as shown in Fig. 2-6. This output impedance is capacitive, and its phase angle is 45 deg at the −3-dB cutoff frequency.

Since the output voltage is taken across the resistor in a high-pass section, and is taken across the capacitor in a low-pass section (Fig. 2-3), the output voltage leads the input voltage for a high-pass section, but the output voltage lags the input voltage for a low-pass section. In accordance with Kirchhoff's voltage law, the sum of the leading voltage across the resistor and the lagging voltage across the capacitor is equal to the input voltage, and is oppositely polarized. In other words, the net sum of the voltages around the circuit is equal to zero at any frequency.

Consider the effect of *tolerances* on the output-impedance value. If the resistor has an actual value of 12,000 ohms, and the ca-

34 · Elements of Electronic Circuit Design

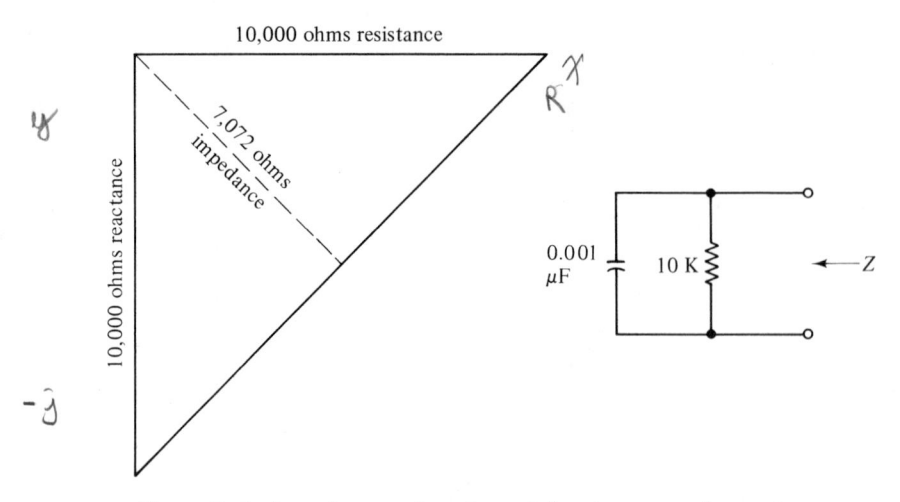

Figure 2–6 Impedance value of parallel resistance and capacitance.

pacitor has an actual value of 0.0012 μF, the output impedance of the *RC* filter section at its cutoff frequency becomes 8500 ohms, as compared with a bogie value of 7072 ohms. This is an error of 20 percent. Next, if the resistor has an actual value of 8000 ohms, and the capacitor has an actual value of 0.0008 μF, the output impedance of the *RC* filter section becomes 5650 ohms, as compared with a bogie value of 7072 ohms. This is an error of 20 percent. Accordingly, both the high-side tolerance and the low-side tolerance limits represent *worst-case* conditions. In summary, the output impedance of the *RC* filter section could range from 5650 ohms to 8500 ohms.

Output Voltage With reference to the high-pass and low-pass *RC* filter sections depicted in Fig. 2–3, the output voltage is less than the input voltage, but it approaches the input-voltage at very high operating frequencies for the high-pass filter, and at very low operating frequencies for the low-pass filter. At any frequency, the ratio of the output voltage to the input voltage is equal to R/Z for the high-pass section, and is equal to X/Z for the low-pass section. This ratio may be found by inspection of the charts shown in Fig. 2–3.

Rolloff Rolloff is defined as the manner in which the amplitude-frequency characteristic of a filter varies as it approaches its frequency limits. It is customary to express rolloff in terms of —dB per octave, or of —dB per decade. With reference to Fig. 2–7, an

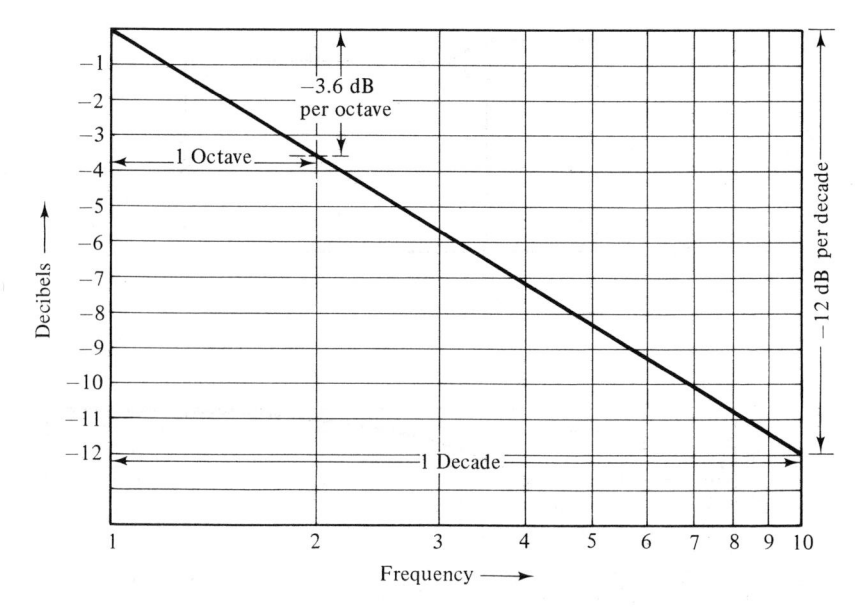

Figure 2-7 Example of dB rolloff per octave and per decade.

octave is defined as the interval over which the operating frequency doubles. A decade is defined as the interval over which the operating frequency increases 10 times. In this example, the output level decreases by 3.6 dB when the frequency is doubled. In other words, the rolloff is equal to −3.6 dB per octave. This is an example of uniform rolloff. The output level decreases by 12 dB when the frequency increases 10 times. Or we say that the rolloff is equal to −12 dB per decade.

With reference to Fig. 2-3, it is observed that the output voltage for an *RC* filter section has an input/output ratio of 1.57 through the −3-dB cutoff point. This is a rolloff rate of approximately −3.9 dB per octave. Note that this is a case of nonuniform rolloff. In other words, if we consider the input/output voltage ratio over the 0.1-to-1, or over the 1.0-to-10 decades for the high-pass and low-pass sections, respectively, this ratio is seen to be 7.07, corresponding to −17 dB per decade. Of course, this is an approximation, because the rolloff is not linear over the decade interval. However, this approximation is acceptable for many circuit-design applications.

Transient Response An *RC* filter has a transient response to a step function of voltage as exemplified in Fig. 2-8. Its response is

36 · Elements of Electronic Circuit Design

(a)

A. — Capacitor voltage on charge.
B. — Capacitor voltage on discharge.
Resistor voltage on charge or discharge.

Time in $R \times C$

(b)

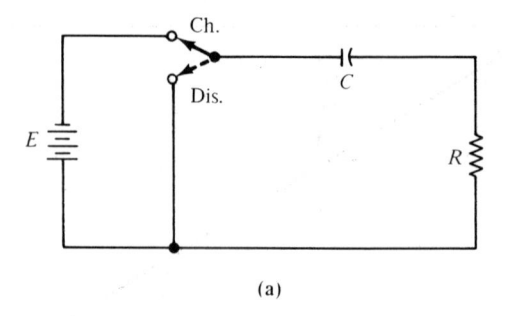

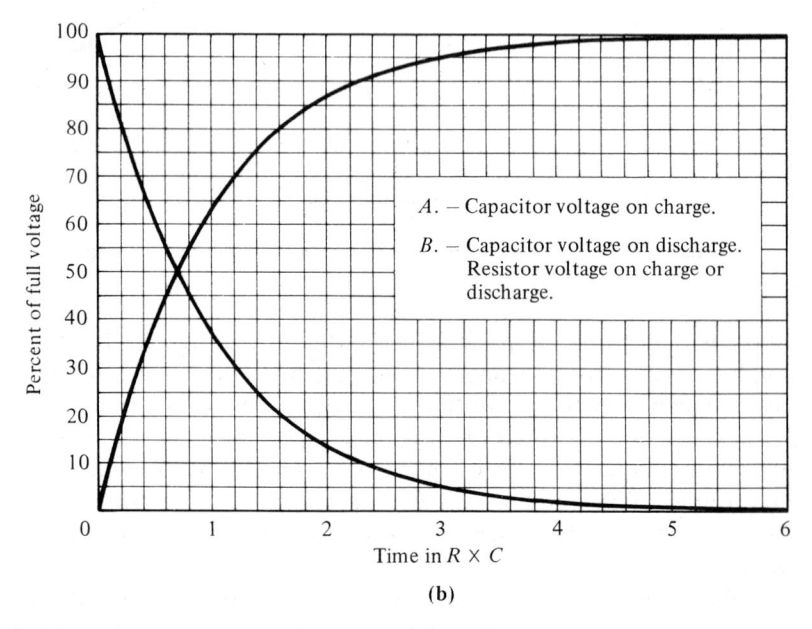

Figure 2–8 Basic transient response; (a) *RC* circuit arrangement; (b) universal time constant chart.

expressed by exponential functions. All *RC* series circuits have the same response waveforms, although the rate of capacitor charge and discharge depends upon the time constant of the circuit. The time constant is defined as the product of *R* and *C;* when *R* is expressed in ohms, and *C* is expressed in farads, their *RC* product is in seconds. For example, the product of one megohm and one microfarad is equal to one second. As seen from Fig. 2–8, the voltage across the capacitor rises to approximately 63 percent of maximum in one time constant; conversely, the capacitor will discharge from its initial

Normalization Technique of Tolerance Analysis · 37

potential to approximately 37 percent of this value in one time constant. If the output voltage is taken across the resistor, the arrangement is called a *differentiating* circuit. If the output voltage is taken across the capacitor, the arrangement is called an integrating circuit. Note in passing that the output waveforms are only rough approximations of true mathematical differentials and integrals. Active filter arrangements must be employed to obtain precise replicas of ideal waveforms.

Consider the effect of component *tolerances* on the time constant of a series RC circuit. If R has a value of 10,000 ohms ±20 percent, and C has a value of 0.001 μF ±20 percent, their RC product (time constant) will fall in the range from 14.4 microseconds to 6.4 microseconds. With respect to the bogie time constant of 10 microseconds, the combination tolerance is plus 44 percent or minus 36 percent. In other words, *the tolerance on the product of two component values is much greater than the tolerance on their sum.*

2–2 NORMALIZATION TECHNIQUE OF TOLERANCE ANALYSIS

Analysis of tolerances is facilitated by normalization of impedance functions, as exemplified in Fig. 2-3. To normalize a function is to standardize the function, or, to place it in a standard form. Here, the percentage of maximum output from the filter section is formulated in terms of the ωRC product. In turn, the normalized function plots as a universal frequency response curve. This mathematical operation is performed as follows:

$$\frac{V_{out}}{V_{in}} = \frac{R}{Z} = \frac{R}{R - jX} = \frac{1}{1 - jX/R}$$

$$\frac{V_{out}}{V_{in}} = \frac{1}{\sqrt{1 + \frac{1}{\omega^2 R^2 C^2}}}$$

Since the output voltage approaches the input voltage as a limit, the output/input ratio ranges from zero to 1. The most useful portion of this range for most applications is a variation in ωRC from 0.01 to 10. Most circuit-design problems are concerned with tolerances on the cutoff frequency ($\omega RC = 1$). It is evident from the normalized form of the impedance function that the worst cases will occur when both R and C have their positive tolerance limits simultaneously, or when they have their negative tolerance limits simultaneously. Thus, if R has its positive tolerance limit when C has its negative tolerance limit, the cutoff frequency error is within its worst-case values.

As another illustration, consider the normalization of an exponential function, as exemplified in Fig. 2–8. Here, the percentage of maximum output from the differentiating (or integrating) section is formulated in terms of the RC product. In turn, the normalized function plots as a universal time-constant chart. This mathematical operation is performed as follows:

$$\frac{V_{out}}{V_{in}} = \epsilon^{-t/RC} \qquad \text{(capacitor charging)}$$

$$\frac{V_{out}}{V_{in}} = (1 - \epsilon^{-t/RC}) \qquad \text{(capacitor discharging)}$$

In turn the normalized functions plot as universal time-constant curves in terms of percentage of full voltage versus the RC product. It is evident from the universal curves that the worst cases will occur when both R and C have their positive tolerance limits simultaneously, or when they have their negative tolerance limits simultaneously.

2–3 OPERATION OF RESISTORS AT HIGH FREQUENCIES

Composition resistors with high values of resistance generally have poor high-frequency characteristics. For example, a typical variation in the effective resistance of a composition resistor versus frequency is shown in Fig. 2–9. At 5 MHz, the effective value of resistance is only a small fraction of its low-frequency effective value. This decrease in effective value is caused chiefly by stray capacitance associated with the design of the resistor. On the other hand, specialized types of resistors are available for high-frequency circuit application. These specialized types require minimum derating of their bogie value in high-frequency operation. Note that a *derating factor* is the factor by which the rating of a component part (or device) is reduced in an application for which its bogie rating is unrealistic.

Note in passing that waveforms with discontinuities have extremely high frequency components. For example, when the switch

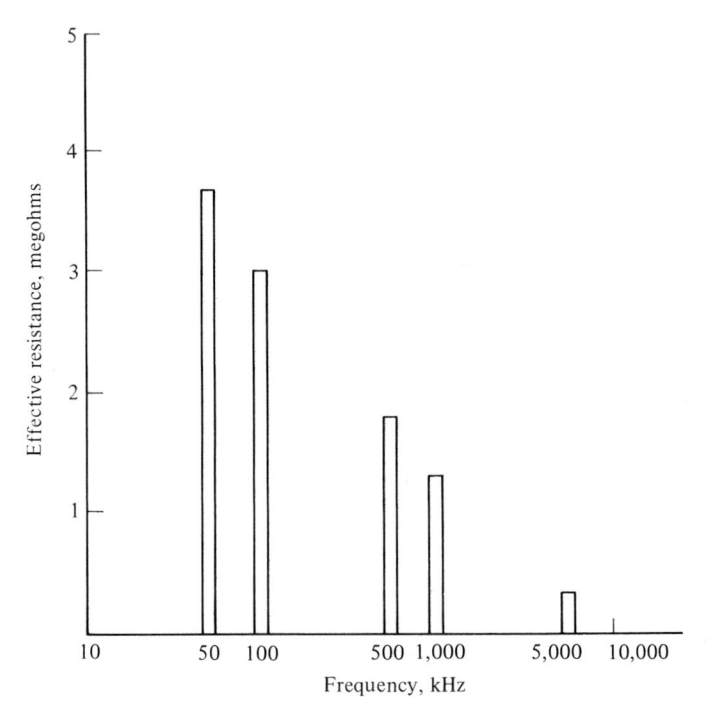

Figure 2-9 Typical variation in effective resistance of a composition resistor versus frequency.

is thrown in Fig. 2-8, and the battery voltage is suddenly applied to the *RC* section, the step function of voltage that is generated rises with extreme rapidity from zero to level *E*. This sudden discontinuity, or voltage change, represents *all* frequencies, up to the highest frequency that can pass without undue attenuation through the stray inductance of the circuit.

2-4 CASCADED *RC* CIRCUITRY

A common cascaded *RC* configuration is depicted in Fig. 2-10. It is assumed in this example that the device has a gain of unity, and that its sole function is to provide isolation between the first and second *RC* sections. In other words, the first section is not loaded by the second section. Consequently, the first section has the frequency characteristic shown in Fig. 2-3(a). Similarly, the second section in

40 · Elements of Electronic Circuit Design

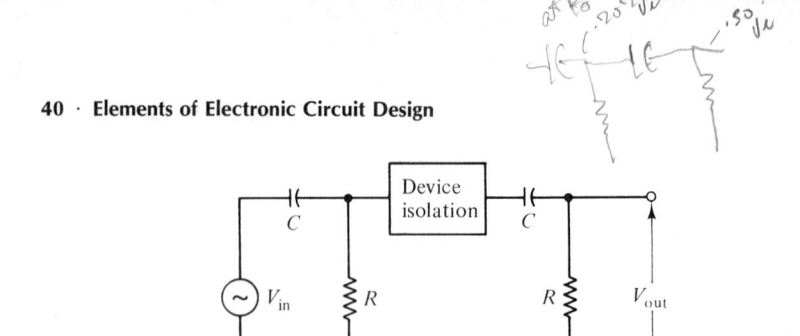

Note: R = R
C = C

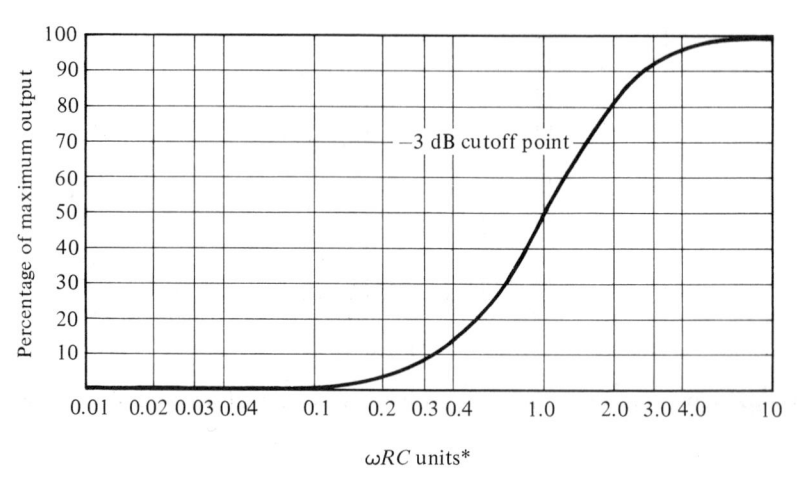

(a)

*R and C units are those utilized in a single section.

(b)

Figure 2–10 Two cascaded RC high-pass sections with device isolation: (a) configuration; (b) universal frequency response chart.

Fig. 2–10 has this same frequency characteristic. From the viewpoint of input/output relations, the output amplitude is equal to the square of the input amplitude. For example, the output from the first section is 0.707 of the input amplitude at the frequency of section cutoff ($\omega RC = 1$). Next, the second section has an output amplitude that is equal to 0.707 of its input amplitude. In other words, the output amplitude from the second section is equal to 50 percent of the input amplitude to the first section, at $\omega RC = 1$.

A universal frequency response chart for the two cascaded sections is shown in Fig. 2–10(b). Note that the rolloff is appreciably steeper than for a single RC section. That is, the rolloff through the −3-dB cutoff point is 5 dB per octave for the two-section arrangement, compared with a rolloff of 3.9 dB per octave for a single section. Observe also that the cutoff frequency is higher for a two-section configuration than for a single-section arrangement. Thus, the −3-dB cutoff frequency in Fig. 2–10(b) corresponds to $\omega RC = 1.55$, approximately, compared with $\omega RC = 1$ for a single section. This change in cutoff frequency is an aspect of the steeper rolloff for the two-section configuration. It is evident that a three-section arrangement would have a still higher −3-dB cutoff frequency.

Consider the effect of component tolerances on the cutoff frequency of a two-section RC filter with device isolation. Because of the comparatively steep rolloff of the two-section configuration, it will be seen that component tolerances have a lesser effect on cutoff-frequency shift than is the case in a one-section arrangement. With reference to Fig. 2–4, observe the −3-dB cutoff frequencies that will result when this RC section is cascaded with a similar section (device isolation included). It is evident that the two-section cutoff frequencies will occur at the 84.1 percent level with respect to the single-section characteristic. Thus, the bogie cutoff frequency for two sections is approximately 7500 Hz. With +20 percent limit tolerances on R and C, the cutoff frequency becomes about 5800 Hz. Or, with −20 percent limit tolerances, the cutoff frequency becomes 11,000 Hz, approximately (*worst-case condition*). This is a cutoff frequency shift of 46 percent, compared with a shift of 56 percent for a single section, as noted previously.

2–5 BASIC GRAPHICAL SOLUTIONS

Vector diagrams on graph paper can frequently provide sufficiently accurate answers to circuit problems in particular applications, and can save time for the circuit designer. With reference to Fig. 2–11, a simple construction permits the designer to determine the parallel impedance of a resistor and a capacitor, from the vector diagram of the series impedance of the same resistor and capacitor at a given operating frequency. Thus, R and X_c combine in series to form the impedance Z_s. On the other hand, if these same values of R and X_c are connected in parallel, their impedance Z_p is given by the altitude of the right triangle.

Conversely, this construction can be reversed to determine the series impedance of a resistor and capacitor, from their parallel im-

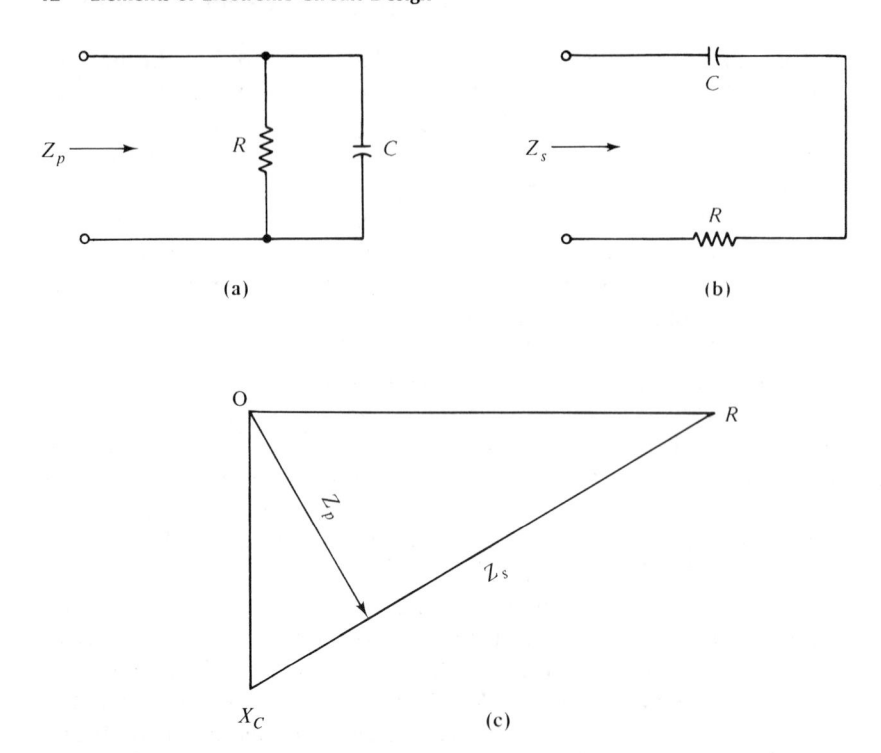

Figure 2-11 Equivalent series and parallel RC sections: **(a)** parallel RC section; **(b)** series RC section; **(c)** equivalent series and parallel resistance and reactance values.

pedance vector. In other words, given the magnitude of Z_p and its phase angle, a right triangle is constructed with Z_p as its altitude. This right triangle then gives the magnitudes of the series impedance and its resistive and reactive components, as in Fig. 2-11(c). The parallel impedance vector Z_p, of course, has rectangular components, as shown in Fig. 2-12. This is a very useful construction, because it can be made with comparative ease, and it shows at a glance how a resistor must be *derated* for effective value when it is paralleled by a significant amount of stray capacitance.

The circuit designer will also find the construction in Fig. 2-12 useful for quick determination of *equivalent series and parallel RC circuits*. In other words, the impedance of R and X_c connected in parallel is the same as the impedance of R_{comp} and $X_{c_{\text{comp}}}$ connected in series. Of course, the equivalence holds only at the frequency that is chosen for evaluation of X_c. The equivalent circuits have the same

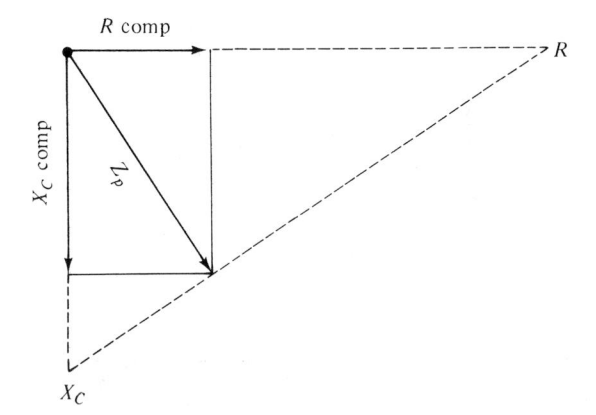

Figure 2-12 Rectangular components of Z_P.

phase angle (same power factor), and dissipate the same amount of power for a given value of applied voltage.

2-6 BASIC *RC* BANDPASS FILTER ACTION

An *RC* bandpass filter is formed by connecting a high-pass section in cascade with a low-pass section. An example of a simple *RC* bandpass filter arrangement is shown in Fig. 2-13. Component values are chosen to provide a −3-dB low-frequency cutoff of 100 Hz, and a high-frequency cutoff of 10 kHz, approximately. The low-end rolloff and the high-end rolloff and the cutoff frequencies follow from the relations shown in Fig. 2-3. From a practical viewpoint, *tolerances* on component values in the bandpass-filter arrangement have the same effect on frequency response as in their low-pass and high-pass sections. There is comparatively little *interaction* between the high-pass and low-pass sections in the example of Fig. 2-13, because the ratio of the cutoff frequencies is 100 to 1. On the other hand, if the circuit designer chooses a cutoff-frequency ratio of 10 to 1, or less, circuit interaction must be taken into account.

When the cutoff-frequency ratio of an *RC* bandpass filter such as that depicted in Fig. 2-13(a) is chosen at 10 to 1, or less, the configuration can no longer be analyzed accurately in terms of its individual high-pass and low-pass section responses. Instead, the bandpass filter must now be treated as a series-parallel configuration over its entire response range. In turn, the response calculations

44 · Elements of Electronic Circuit Design

(a)

(b)

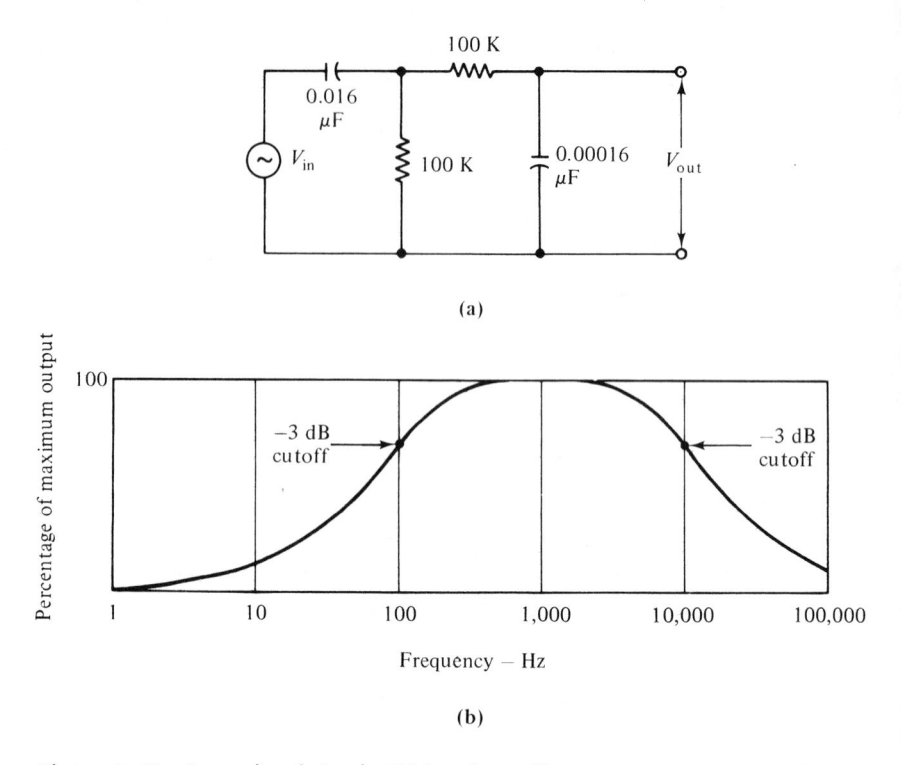

Figure 2-13 Example of simple RC bandpass filter arrangement: **(a)** configuration; **(b)** frequency response curve.

become comparatively involved. On the other hand, the bandpass filter response is easily analyzed from the charts of Fig. 2-3 if device isolation is included between the high-pass and low-pass sections, as depicted in Fig. 2-14. Thus, if a device gain of unity is assumed, the universal-chart data are directly applicable, regardless of the chosen cutoff-frequency ratio.

Consider the *insertion loss* of the bandpass-filter arrangement in Fig. 2-14, if the designer chooses the same cutoff frequency for the high-pass section and the low-pass section. If a device gain of unity is assumed, as before, it is evident that each section will introduce a 3-dB loss at the midband frequency of the filter, inasmuch as each section operates at its cutoff frequency. In turn, the insertion loss of the bandpass filter is 6 dB in this example. Note in passing that the insertion loss denotes the maximum available V_{out}/V_{in} value. Observe also that *component tolerances* in the configuration of

RC Band Elimination Filter · 45

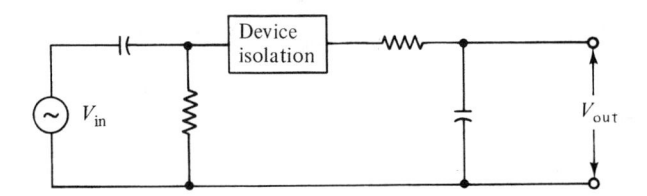

Figure 2-14 Basic RC bandpass filter arrangement with device isolation.

Fig. 2-14 have the same effect on bandpass filter response as on their individual section responses. If the high-pass and low-pass responses overlap appreciably, tolerances on both sections must be included in calculation of the resultant midband response.

When comparatively rapid rolloff is desired in the frequency response of an *RC* bandpass filter, multisection high-pass and low-pass configurations can be employed, as exemplified in Fig. 2-15. Note that when a pair of high-pass *RC* sections are connected in cascade, the configuration must be analyzed as a series-parallel arrangement. Similarly, when a pair of low-pass *RC* sections are connected in cascade, the configuration must be analyzed as a series-parallel arrangement. This requirement results from the loading that the second section places on the first section. Note also that when two *RC* sections are connected in cascade, the cutoff frequency of the combination is shifted, in somewhat the same manner as was explained for the configuration in Fig. 2-10. For example, if one *RC* section has a cutoff frequency of 1900 Hz, and it is connected to a following section that has the same cutoff frequency, then the cutoff frequency for the two-section low-pass filter becomes approximately 800 Hz.

2-7 RC BAND ELIMINATION FILTER

A widely used type of band-elimination or notch filter is the *RC* parallel-T configuration, shown in Fig. 2-16. It is essentially a low-pass *RC* tee section paralleled with a high-pass *RC* tee section. The basic filter uses component values such that a *null* or zero output occurs at a specified frequency. Zero output is the result of equal output amplitudes with 180-deg phase difference from the two *RC* sections. A basic configuration is shown in Fig. 2-16. The shunt resistor has half the value of either of the series resistors, and the shunt capacitor has double the value of either of the series capaci-

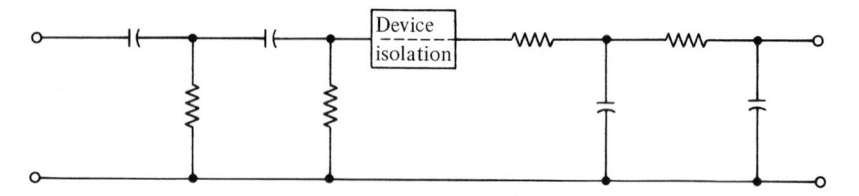

Figure 2-15 Bandpass RC filter arrangement with double-section input and double-section output.

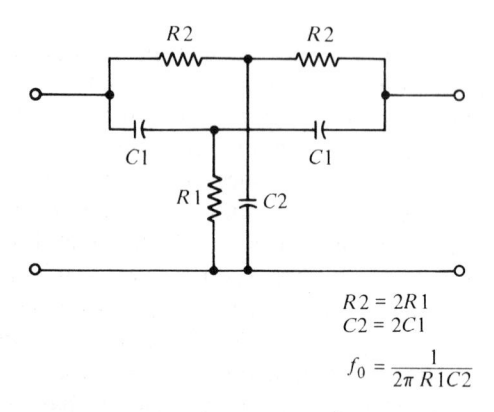

$$R2 = 2R1$$
$$C2 = 2C1$$
$$f_0 = \frac{1}{2\pi R1 C2}$$

Figure 2-16 Basic parallel-T RC band-elimination filter.

tors. If close-tolerance components are used, a null output is obtained at frequency f_0. Note that a pure sine-wave input is required to obtain a true null output. In other words, if the input waveform contains harmonics, only a shallow (incomplete) null will be obtained at f_0.

An example of a frequency response curve for a parallel-T RC filter section is shown in Fig. 2-17. The null frequency is 4.4 kHz, approximately, and the bandwidth is about 4.9 kHz. Consider the effect of component *tolerances* on the null frequency. When all tolerances are at the lower limit, the *worst-case* condition results. Thus, if all component values are 20 percent low, the null frequency becomes 56 percent higher than bogie. On the other hand, if all component values are 20 percent high, the null frequency becomes 31 percent lower than bogie. Note that a complete null is still obtained when all tolerances are at their upper limit, or when they are at their lower limit, because the $R1/R2$ and $C1/C2$ ratios remain the same.

Transient Response of Multisection RC Filters · 47

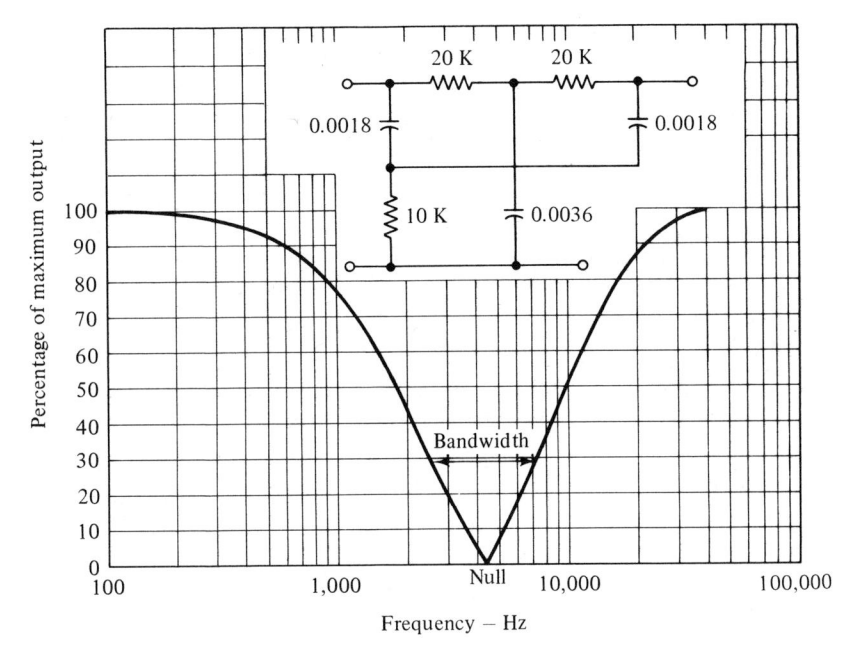

Figure 2-17 Example of a frequency response curve for a parallel-T RC filter section.

2−8 TRANSIENT RESPONSE OF MULTISECTION RC FILTERS

Multisection RC filters are used as integrators in various applications. A universal time-constant chart for one-, two-, and three-section symmetrical integrators (low-pass filters) is shown in Fig. 2–18. It is seen that the output is slowed, or delayed, as more sections are added to an RC filter. Note that the shape of the output waveform is changed as more sections are added. Curve A is an exponential waveform. However, curve B is a modified exponential waveform. Curve C is a further modified exponential waveform. It is interesting to note that when an unlimited number of low-pass RC filter sections are connected in cascade, the output waveform produced by a step-function input approaches a Gaussian distribution curve in its shape. This characteristic is of basic interest to the designer of oscilloscope amplifiers for complex waveforms.

Component tolerances have the same general effect on multi-section filter operation as on single-section operation. Insofar as time-constant change is concerned, the *worst-case condition* occurs

48 · Elements of Electronic Circuit Design

(a) One-section integrator.

(b) Two-section integrator. Note: $R = R$, $C = C$

(c) Three-section integrator.

Note: $R = R = R$, $C = C$

(d) Time-constant chart

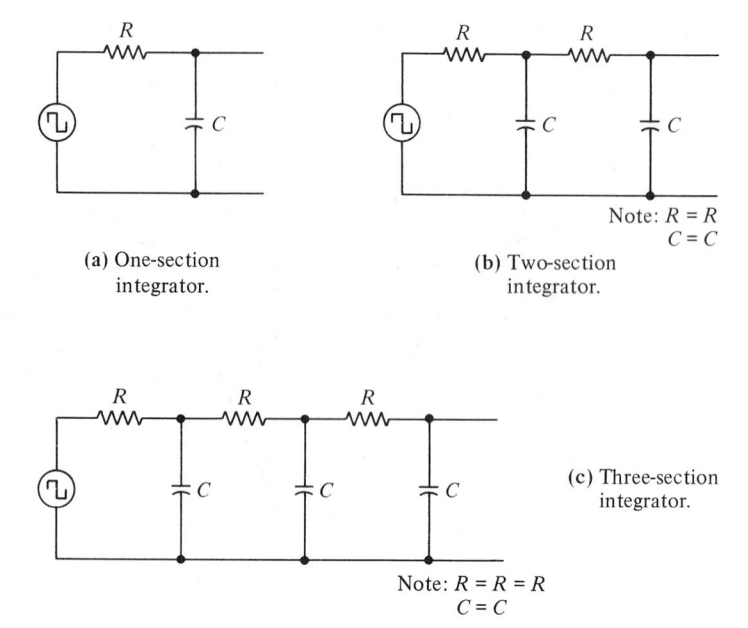

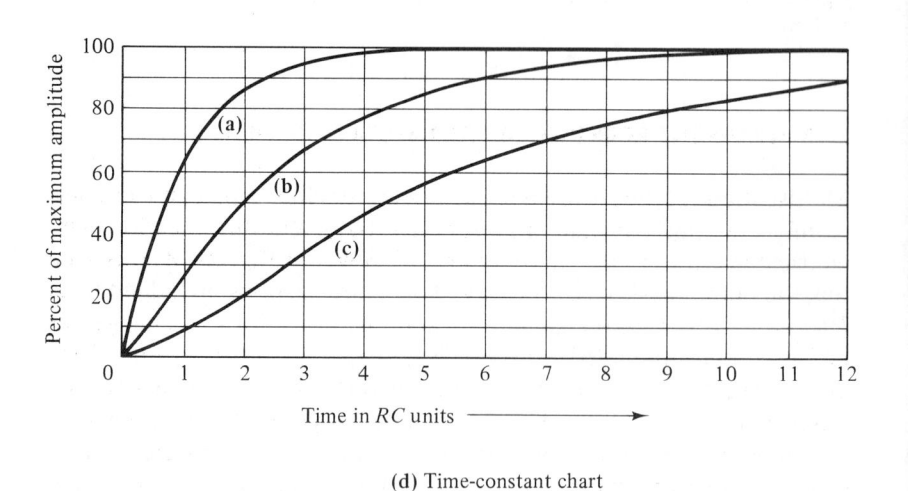

Figure 2–18 Universal time constant chart for one-, two-, and three-section integrator circuits.

when both R and C have their positive-tolerance limit values. This time-constant error is 44 percent for a single section, if R and C

both have 20 percent tolerances. Again, if both R and C have their negative-tolerance limit values, the time-constant error is 36 percent for a single section. Consider the worst-case condition for a two-section configuration. A bogie time constant of 1 for a single section becomes an erroneous time constant of 1.44. With positive-limit values in the second section also, the filter remains symmetrical, and the effective time constant of the two-section configuration becomes 4.3 × 1.44, or 6.2 as compared with 4.3 for a bogie two-section configuration. This is a time-constant error of 44 percent for the two-section arrangement.

2–9 MONTE CARLO METHOD

A Monte Carlo method denotes any procedure that involves statistical sampling techniques in order to obtain a probabilistic approximation to the solution of a mathematical or physical problem. Thus, this method is an aspect of *quality control*. The practice of quality control denotes the control of variation in workmanship, processes, and materials, in order to produce a consistent, uniform product. Sampling techniques are often utilized when 100 percent product testing would be prohibitively expensive. The statistical aspect involves the trend of test results. In other words, if a particular sampling rate yields a finding of few or no rejects, the director of engineering may decide to reduce the sampling rate in order to lessen production costs. On the other hand, if the sampling rate yields a finding of many rejects, a production change order will ordinarily be issued, and the sampling rate will be increased. Then, if it becomes evident that the change order has solved the reject problem satisfactorily, the sampling rate would ordinarily be reduced.

2–10 CIRCUIT DESIGN BREADBOARD

Circuit designers use various kinds of breadboards for temporary interconnection of components and devices for performance tests and measurements. A breadboard model comprises a rough assembly form that serves to prove the feasibility or the faults of a theoretical circuit or system. A typical solderless breadboarding unit is illustrated in Fig. 2–19. It contains a (59 × 10)-hole matrix. Component or device leads and stripped connecting wires are plugged into the holes for interconnection into a circuit. The matrix is arranged in

Figure 2-19 A typical circuit designer breadboard. *(Courtesy, Continental Specialties Corp.)*

two 59 × 5 sets, with each of the 59 columns provided with five electrically common contacts that afford multiple connection points for each component or device lead. The column-to-column and row-to-row hole spacing accommodates standard dual in-line package (DIP) integrated circuits (IC's). Above and below the main solderless "socket" are bus strips, each provided with 50 holes.

CHAPTER 3

Fundamentals of Wave Filter Design

3-1 GENERAL CONSIDERATIONS

Although any filter arrangement may be technically described as a *wave filter,* circuit designers generally limit this term to LC filter sections and networks in which resistive components are minimized. Because wave filters consist of various combinations of inductors and capacitors, they have a superficial resemblance to resonant (tuned) circuits. Although a wave filter appears to be a configuration of series-resonant and parallel-resonant circuits, a filter arrangement generally has a frequency response that is unexpected from the viewpoint of elementary series- and parallel-resonant circuit response. In other words, a wave filter does not have a resonant frequency. Instead, a wave filter typically passes all frequencies up to its cutoff point, and rejects all frequencies above its cutoff point. For example, a video-frequency filter typically passes all frequencies from zero frequency (DC) to 4 MHz; in turn, this filter rejects all frequencies above 4 MHz. As another example, a UHF filter rejects all frequencies below 471 MHz, and passes all frequencies above 471 MHz.

It is instructive to consider the simplest form of an L-section low-pass wave-filter circuit, shown in Fig. 3–1. It consists of a series inductor and a shunt capacitor. If we disregard the load resistor R_L, L and C operate as an ordinary series-resonant circuit. However, with a suitable value of R_L connected across the output terminals of the L section, the series-resonant circuit then functions as a wave fil-

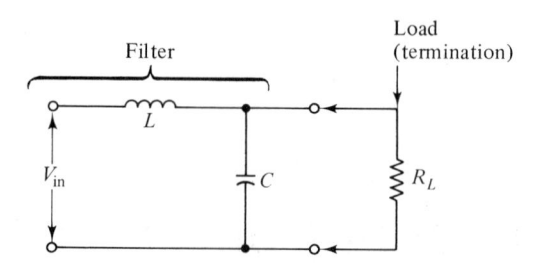

Figure 3-1 Simplest L-section low-pass wave-filter circuit.

ter. As such, all frequencies up to a certain *cutoff frequency* are passed, and all frequencies beyond the cutoff frequency are attenuated or rejected. Some fundamental characteristics of this filter section are evident:

1. No attenuation is imposed at zero frequency (DC), if the winding resistance of L can be ignored, as is usually the case.
2. Complete rejection of extremely high frequencies occurs, because the reactance of L becomes virtually infinite.
3. There will be a frequency at which the output voltage is 3 dB down from its zero-frequency value; this is the *cutoff frequency* of the section.

From previous discussion of RC filters, it would be expected that if a number of LC sections are connected in cascade, the rolloff through the cutoff frequency would become comparatively steep. This is true. Another basic characteristic of cascaded L sections is called the *characteristic impedance* (or *characteristic resistance*) of an infinite series. This concept is developed in Fig. 3-2 for the simple case of a resistive ladder, consisting of a series of cascaded resistive L sections. As successive sections are included in the ladder, the input resistance approaches a limit of 37 ohms in this example. This limiting value of input resistance is called the characteristic resistance of the infinite ladder. In other words, if R_{AB} in Fig. 3-2(a) had a value of 37 ohms, the E/I ratio of the source would be the same as in Fig. 3-2(h).

Next, it is helpful to consider the significance of this characteristic resistance with respect to a single L section. In Fig. 3-3, we have terminated a single L section in its characteristic resistance.

General Considerations · 53

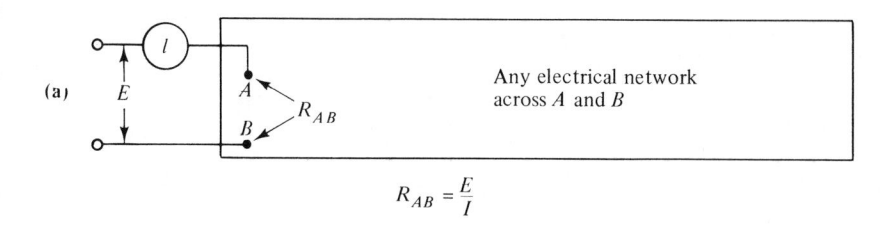

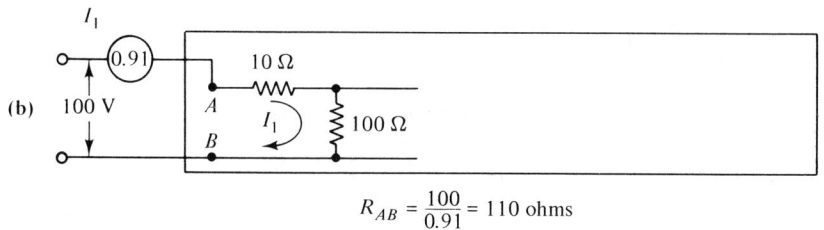

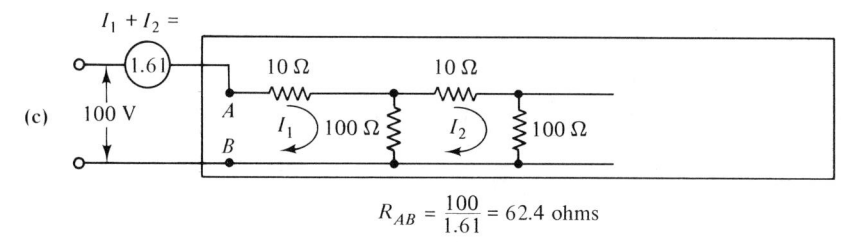

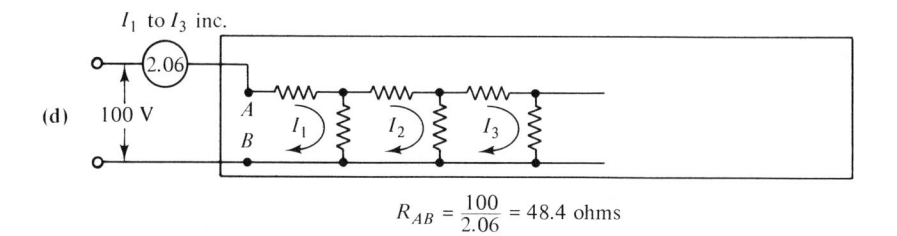

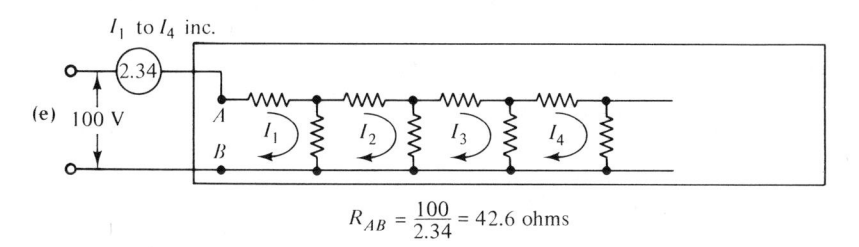

(a)
$$R_{AB} = \frac{E}{I}$$

Any electrical network across A and B

(b)
$$R_{AB} = \frac{100}{0.91} = 110 \text{ ohms}$$

(c)
$$R_{AB} = \frac{100}{1.61} = 62.4 \text{ ohms}$$

(d) I_1 to I_3 inc.
$$R_{AB} = \frac{100}{2.06} = 48.4 \text{ ohms}$$

(e) I_1 to I_4 inc.
$$R_{AB} = \frac{100}{2.34} = 42.6 \text{ ohms}$$

Figure 3–2 Development of the characteristic resistance concept.

54 · Fundamentals of Wave Filter Design

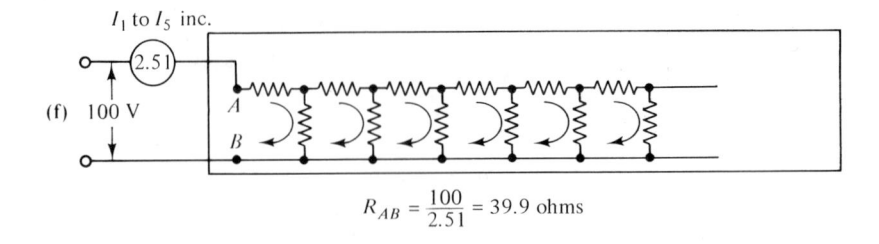

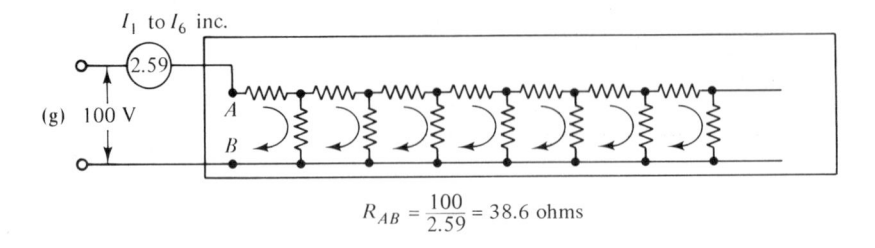

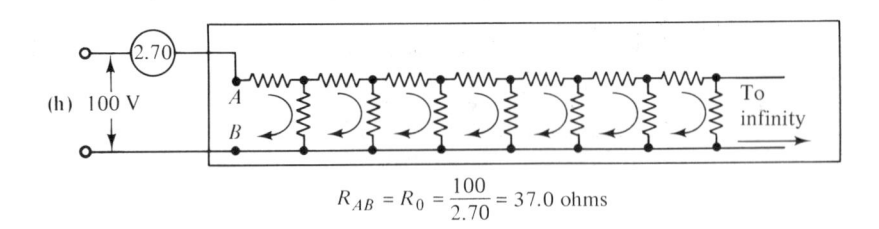

(f) 100 V, I_1 to I_5 inc., 2.51

$$R_{AB} = \frac{100}{2.51} = 39.9 \text{ ohms}$$

(g) 100 V, I_1 to I_6 inc., 2.59

$$R_{AB} = \frac{100}{2.59} = 38.6 \text{ ohms}$$

(h) 100 V, 2.70, To infinity

$$R_{AB} = R_0 = \frac{100}{2.70} = 37.0 \text{ ohms}$$

Figure 3–2 *(Cont.)*

Thus, the parallel combination of 100 ohms and 37 ohms is equal to 27 ohms. Or the input resistance to the *L* section is 37 ohms — the same value as the terminating resistance. Therefore, we say that this *L* section has a characteristic resistance of 37 ohms. Next, consider the effect of *resistive tolerances* on the characteristic resistance of this *L* section. If the series and shunt resistors have a tolerance of ±20 percent each, the characteristic resistance can range from 30 ohms to 44.4 ohms. The *worst-case error* on the characteristic resistance is 20 percent.

An *LC* wave filter also has a certain value of characteristic impedance within its pass band. This impedance value increases past the cutoff frequency of the filter. In turn, the filter attenuates or

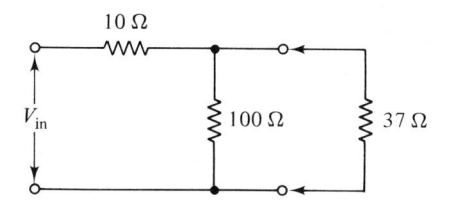

Figure 3-3 A properly terminated resistive L section.

blocks signal passage in the range past its cutoff frequency. With reference to Fig. 3-1, the cutoff frequency of this elementary low-pass filter is given by the approximate equation:

$$f_c \approx \frac{1}{\pi \sqrt{LC}}$$

Note: The symbol $\approx$ means "is approximately equal to."

Thus, if L has a value of 50 μH, and C has a value of 500 pF, the cutoff frequency of the low-pass filter section in Fig. 3-1 will be approximately 2 MHz when the section is properly terminated. Note that the resonant frequency of L and C is equal to 1 MHz. The section is properly terminated by a load of 316 ohms. In other words, the characteristic impedance of the filter section is given by the approximate equation:

$$Z \approx \sqrt{\frac{L}{C}}$$

Consider the effect of component tolerances on the cutoff frequency and on the characteristic impedance of the foregoing low-pass filter section. The cutoff frequency is inversely proportional to the square root of the LC product. In turn, if both L and C have ±20 percent tolerances, the *worst-case* error in cutoff frequency will be 80 percent. Again, the characteristic impedance is defined as the square root of the L/C ratio. The worst-case condition occurs when L has its positive-tolerance limit value and C has its negative-tolerance limit value. In turn, the worst-case error in characteristic impedance is 22 percent.

3–2 MULTISECTION *LC* WAVE FILTER

The development of a multisection resistive ladder was shown in Fig. 3–2. A multisection *LC* wave filter can be formed by connecting *L* sections in cascade. In turn, a faster rolloff (sharper cutoff) is obtained. Note that circuit designers generally prefer to form multisection filters from pi sections or from *T* sections, because the calculations are less involved. Basic low-pass *LC* pi and *T* sections are depicted in Fig. 3–4. Observe that each inductor in the *T* configuration is stipulated to have a value of $L/2$. Again, each capacitor in the pi configuration is stipulated to have a value of $C/2$. This convention is adopted so that the foregoing approximate equations for the cutoff frequency and the characteristic impedance of an *L* section will also be applicable to a *T* section or to a pi section. Observe that R_G denotes the internal resistance of the generator. Generally, R_G is equal to R_L, and R_L is equal to the characteristic impedance of the filter.

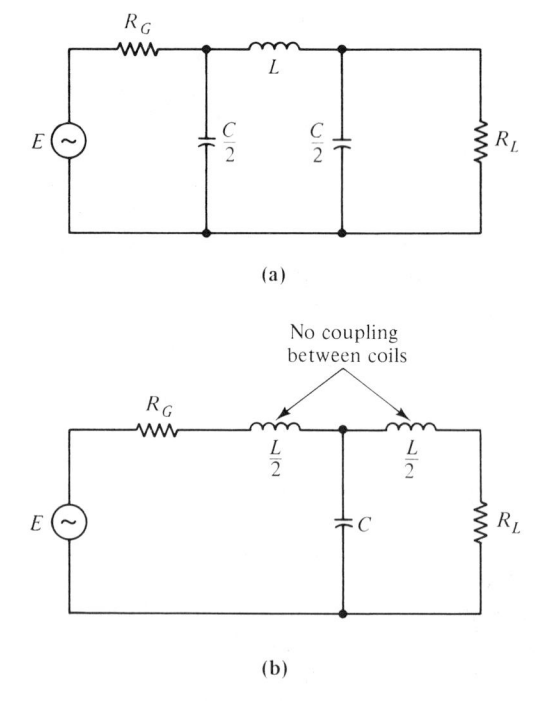

Figure 3–4 Basic symmetrical low-pass *LC* filter sections: **(a)** pi configuration; **(b)** *T* configuration.

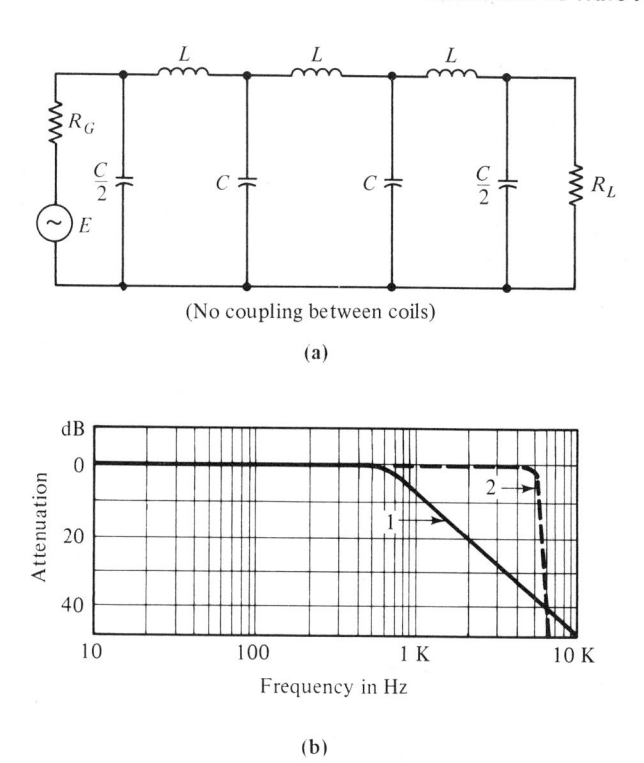

Figure 3–5 Three *LC* low-pass pi sections connected in cascade: **(a)** configuration; **(b)** frequency responses for one-section and three-section filters — (1) response for one section; (2) response for three sections.

Consider the connection of three *LC* low-pass pi sections into a multisection wave filter, as shown in Fig. 3-5. As shown in the frequency response curves, the rolloff for one section is approximately 40 dB per decade, whereas the rolloff for three sections in cascade is 40 dB in about 0.1 decade. Next, observe the basic high-pass *LC* pi and *T* sections depicted in Fig. 3-6. Each inductor in the pi configuration is stipulated to have a value of 2*L*. Again, each capacitor in the *T* configuration is stipulated to have a value of 2*C*. This convention is adopted so that the approximate equations that were previously noted for *L*-section cutoff frequency and characteristic impedance will also be applicable to a pi section or to a *T* section. Note that the rolloff for one section is approximately 40 dB per decade, as shown by curve 2. If the inductor has appreciable resistance (reduced *Q* value), the rolloff becomes slower, as exemplified

58 · Fundamentals of Wave Filter Design

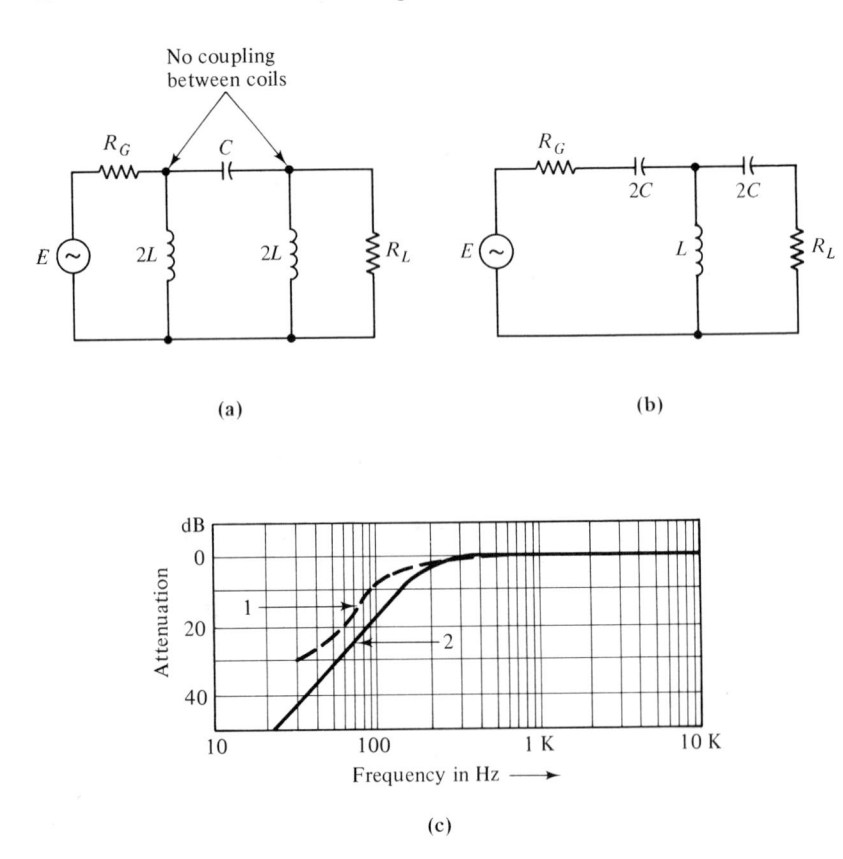

Figure 3–6 Basic high-pass LC pi and T section filters: **(a)** pi section configuration; **(b)** T section configuration; **(c)** frequency response for single section; (2) high-Q inductor; (1): reduced-Q inductor.

by curve 1. When three *LC* high-pass pi sections or *T* sections are connected into a cascaded filter configuration, the rolloff becomes comparatively rapid, with a rate of approximately 40 dB in 0.1 decade.

When the resistance of an inductor is negligible, it presents a purely inductive reactance. Most capacitors present a practically pure capacitive reactance. When such inductors and capacitors are employed in *T*-section or pi-section low-pass or high-pass filters, the product of the series and shunt reactances is a constant that is independent of frequency. In turn, this type of filter is called a *constant-k filter*. Note that the pi and *T* filter sections depicted in Figs. 3–4 and 3–6 are termed *prototype* filter sections. They are the basic configurations from which all types of constant-k filters are developed. It has been pointed out that the simple equations written previously for the cutoff frequency and for the characteristic impedance of a low-pass or high-pass filter are only approximate. However, these equations become more accurate as more sections are connected in cascade. However, an unlimited number of sections would be required in a simple multisection constant-k filter for these equations to become exact.

Composite Wave Filters Because the characteristic impedance of a simple constant-k filter is not independent of frequency, this type of filter can be properly terminated at only one frequency in practice. That is, if the operating frequency is changed, the value of the load would also have to be changed to keep the filter terminated properly. The practical meaning of this fact is seen in Fig. 3-7. Thus, if the prototype high-pass section is properly terminated at 6 kHz, it will be improperly terminated at 3 kHz, and there will be a transmission loss at this frequency. Instead of a zero transmission loss, there will be approximately 7 dB loss, as indicated by the dotted curve. Also, the rolloff is much more gradual than if the prototype section were properly terminated over its entire pass band.

With reference to Fig. 3-7(b), it is seen that if a prototype constant-k low-pass section is properly terminated at zero frequency, it will be improperly terminated at 3 kHz, and there will be a transmission loss at this frequency. Instead of a zero transmission loss, there will be approximately 7 dB loss, as indicated by the dotted curve. Also, the rolloff is much more gradual than if the prototype section were properly terminated over its entire pass band. The practical management of this difficulty exploits the characteristics of *composite filters* that consist of prototype sections supplemented by *m-derived* sections. When an *m-derived end section* is connected at the input and at the output of a constant-k prototype section, the transmission loss through the pass band remains practically zero until the cutoff frequency is approached. Then, the characteristic impedance of the filter increases rapidly and the transmission loss increases rapidly in turn.

Design equations for low-pass constant-k *m*-derived pi sections and T sections are given in Fig. 3-8. Design equations for *m*-derived end sections for use with intermediate pi sections or intermediate T sections are also given. Note that an *m*-derived end section that is designed for use with an intermediate pi section must not be used with an intermediate T section. Similarly, an *m*-derived end section that is designed for use with an intermediate T section must not be

60 · Fundamentals of Wave Filter Design

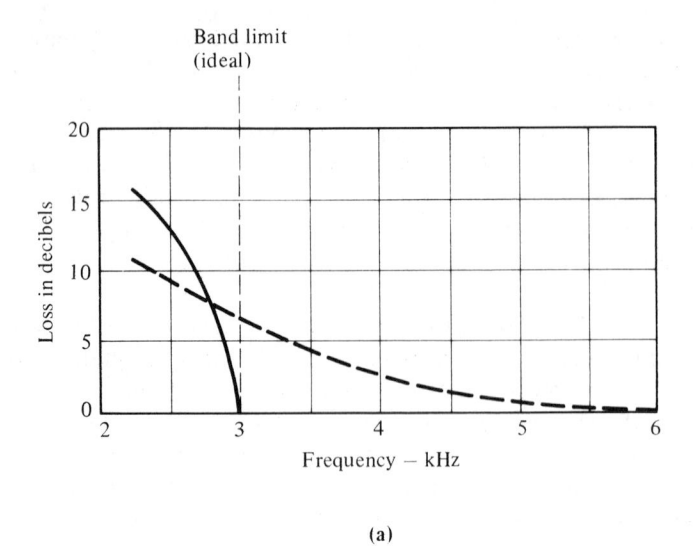

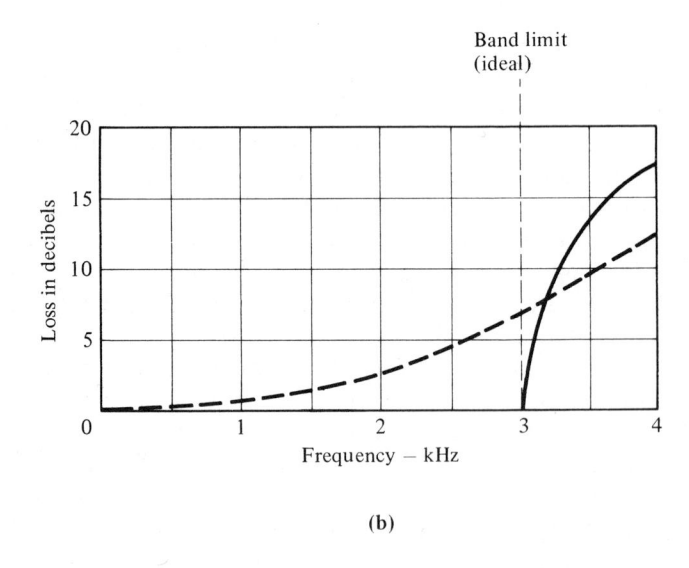

Figure 3–7 Practical effect of frequency-dependent characteristic impedance: **(a)** high-pass section; ideal cutoff, solid curve; practical cutoff, dotted curve; **(b)** low-pass section; ideal cutoff, solid curve; practical cutoff, dotted curve.

Constant-k π section

m–derived π section

Constant-k T section

m–derived T section

$$L_k = \frac{R}{\pi f_c}$$
$$C_k = \frac{1}{\pi f_c R}$$

$$L_1 = mL_k \qquad L_2 = \frac{1-m^2}{4m} L_k$$
$$C_1 = \frac{1-m^2}{4m} C_k \qquad C_2 = m\,C_k$$

m–derived end sections for use with intermediate π section

m–derived end sections for use with intermediate T section

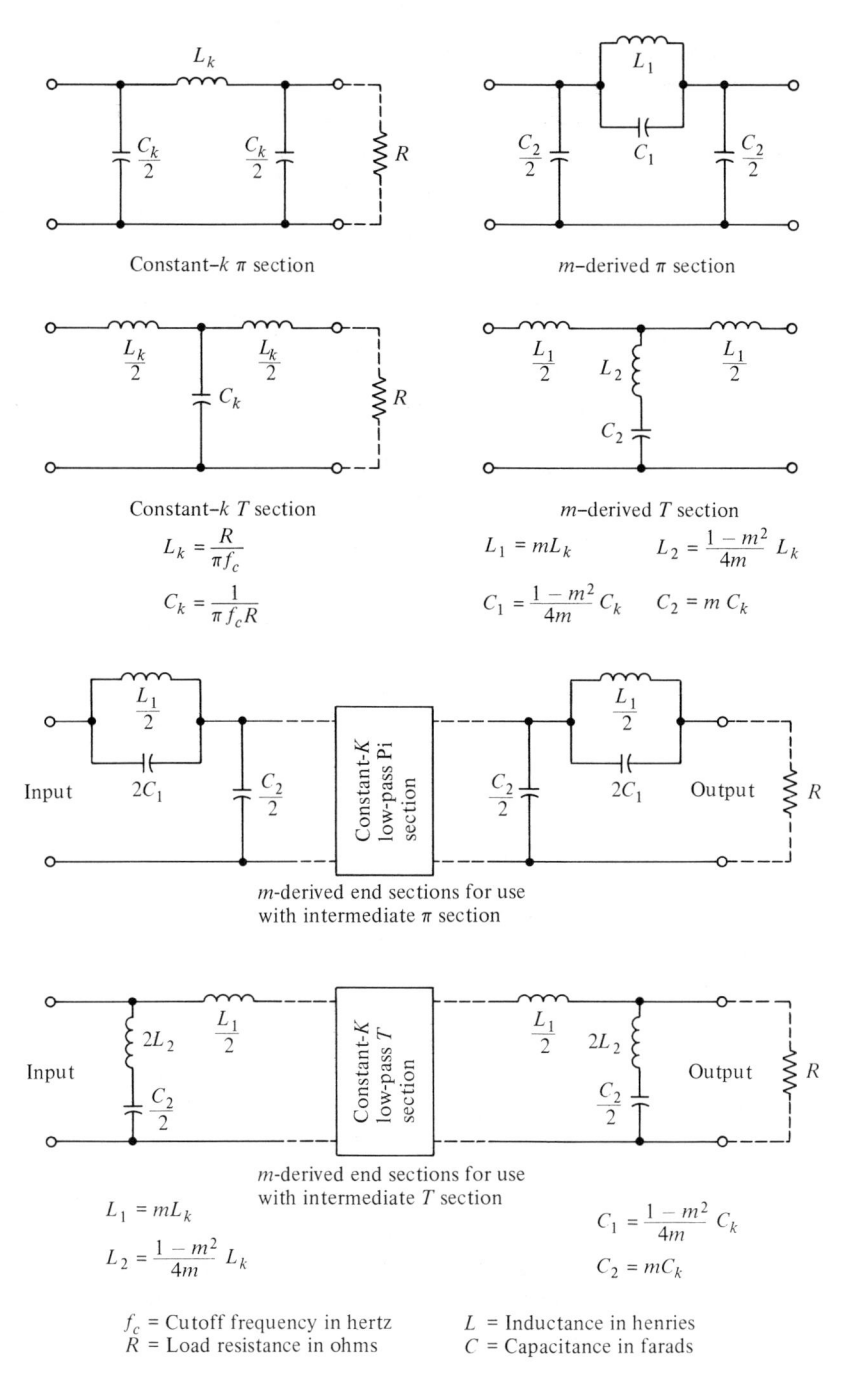

$$L_1 = mL_k$$
$$L_2 = \frac{1-m^2}{4m} L_k$$

$$C_1 = \frac{1-m^2}{4m} C_k$$
$$C_2 = mC_k$$

f_c = Cutoff frequency in hertz
R = Load resistance in ohms
L = Inductance in henries
C = Capacitance in farads

Figure 3–8 Configurations and design equations for m-derived sections and m-derived end sections; low-pass constant-k filters.

used with an intermediate pi section. However, intermediate pi or T sections may be either of prototype constant-k or of m-derived design. Most circuit designers utilize prototype constant-k intermediate sections. In most arrangements, one intermediate section is used; however, several intermediate sections may be included to obtain more nearly ideal filter response. *In the design of m-derived sections, the factor m is generally assigned a value of 0.6 for optimum filter operation.* Note that the filter should be terminated in a pure resistance with a value equal to the characteristic impedance chosen in the design of the sections. Any reactive component in the termination will distort the frequency response of the filter.

Design equations for high-pass constant-k m-derived pi sections and T sections are given in Fig. 3-9. Also shown are design equations for m-derived end sections for use with intermediate pi sections or T sections. Note that when more than one inductor is used in a constant-k filter network, it is essential to avoid any coupling between inductors. Otherwise, the filter characteristics will be disturbed. Inductors may be enclosed in shields, or they may be mounted at right angles to one another in order to eliminate magnetic coupling. Note also that an inductor is susceptible to stray-field pickup, with resulting introduction of interference into the filter system. Therefore, it may be necessary for the circuit designer to enclose an entire filter network in a shield box. Coils of toroidal design are sometimes used to relax coupling and stray-field pickup problems. A toroidal coil is a uniformly wound circular magnetic core.

Next, consider the bandpass LC filter configurations shown in Fig. 3-10. A bandpass filter is an LC network with a purely resistive load that theoretically permits frequencies between two selected cutoff frequencies to pass through with no loss, whereas all frequencies outside of this pass band are rejected. Although no practical design of bandpass filter can have ideal characteristics, the more elaborate networks approach theoretical performance as closely as may be desired. Typical frequency response curves for single-section constant-k bandpass filters are shown in Fig. 3-11. Observe that the rate of rolloff is significantly increased by the use of high-Q inductors. Also, two bandpass sections may be connected in series to increase the rate of rolloff.

Like other prototype filter configurations, the bandpass sections depicted in Fig. 3-10 do not have a constant characteristic impedance through their pass bands. The practical management of this undesired feature exploits the characteristics of composite bandpass filters that consist of a prototype section supplemented by m-derived end sections. With reference to Fig. 3-12, a prototype constant-k filter section

Multisection *LC* Wave Filter · 63

Constant-*k* π section

m-derived π section

Constant-*k* T section

m-derived T section

$$L_k = \frac{R}{4\pi f_c}$$

$$C_k = \frac{1}{4\pi f_c R}$$

$$L_1 = \frac{4m}{1 - m^2} L_k \qquad C_1 = \frac{C_k}{m}$$

$$L_2 = \frac{L_k}{m} \qquad C_2 = \frac{4m}{1 - m^2} C_k$$

m-derived end sections for use with intermediate π section

m-derived end section for use with intermediate T section

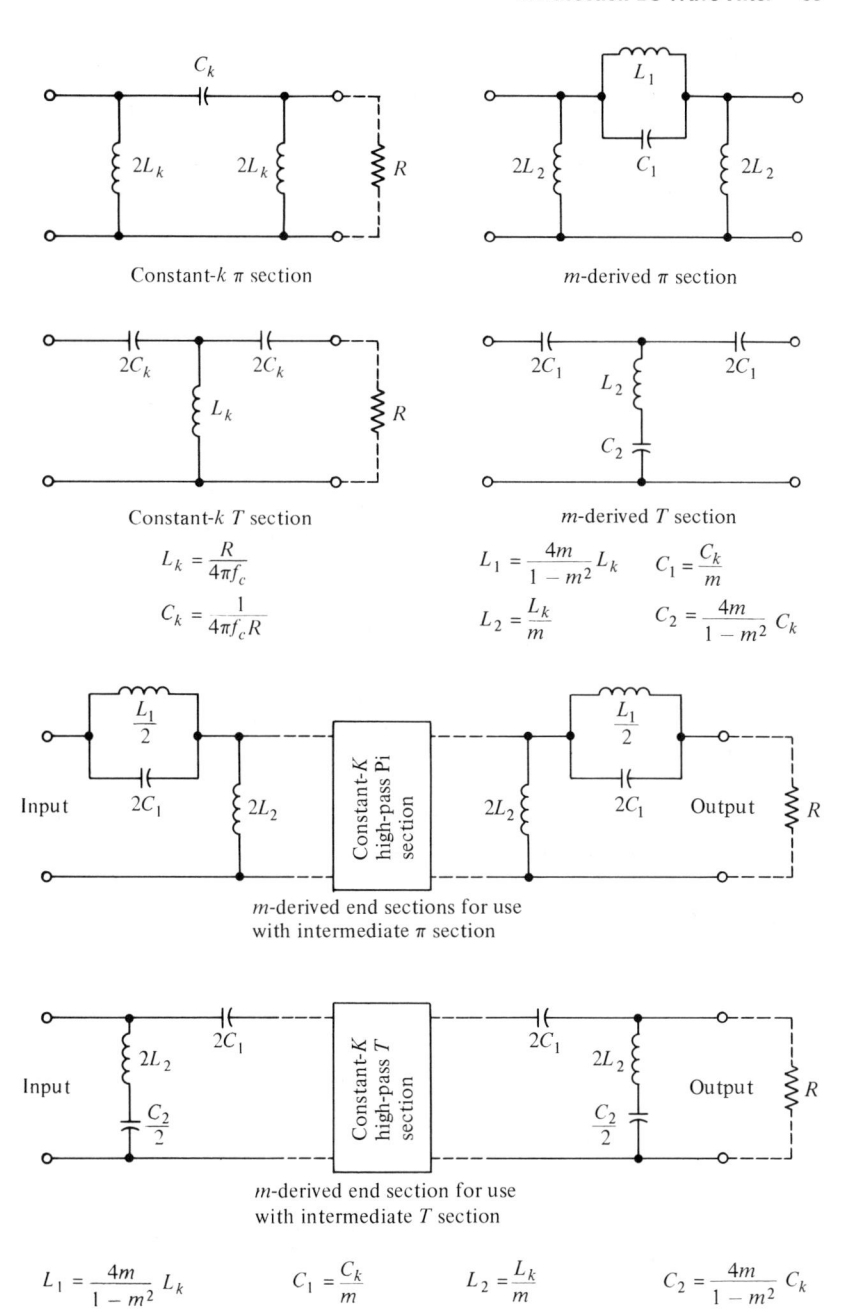

$$L_1 = \frac{4m}{1 - m^2} L_k \qquad C_1 = \frac{C_k}{m} \qquad L_2 = \frac{L_k}{m} \qquad C_2 = \frac{4m}{1 - m^2} C_k$$

Figure 3–9 Configurations and design equations for *m*-derived sections and *m*-derived end sections; high-pass constant-*k* filters.

Constant-k π section

$$L_{1k} = \frac{R}{\pi(f_2 - f_1)} \qquad C_{1k} = \frac{f_2 - f_1}{4\pi f_1 f_2 R}$$

Constant-k T section

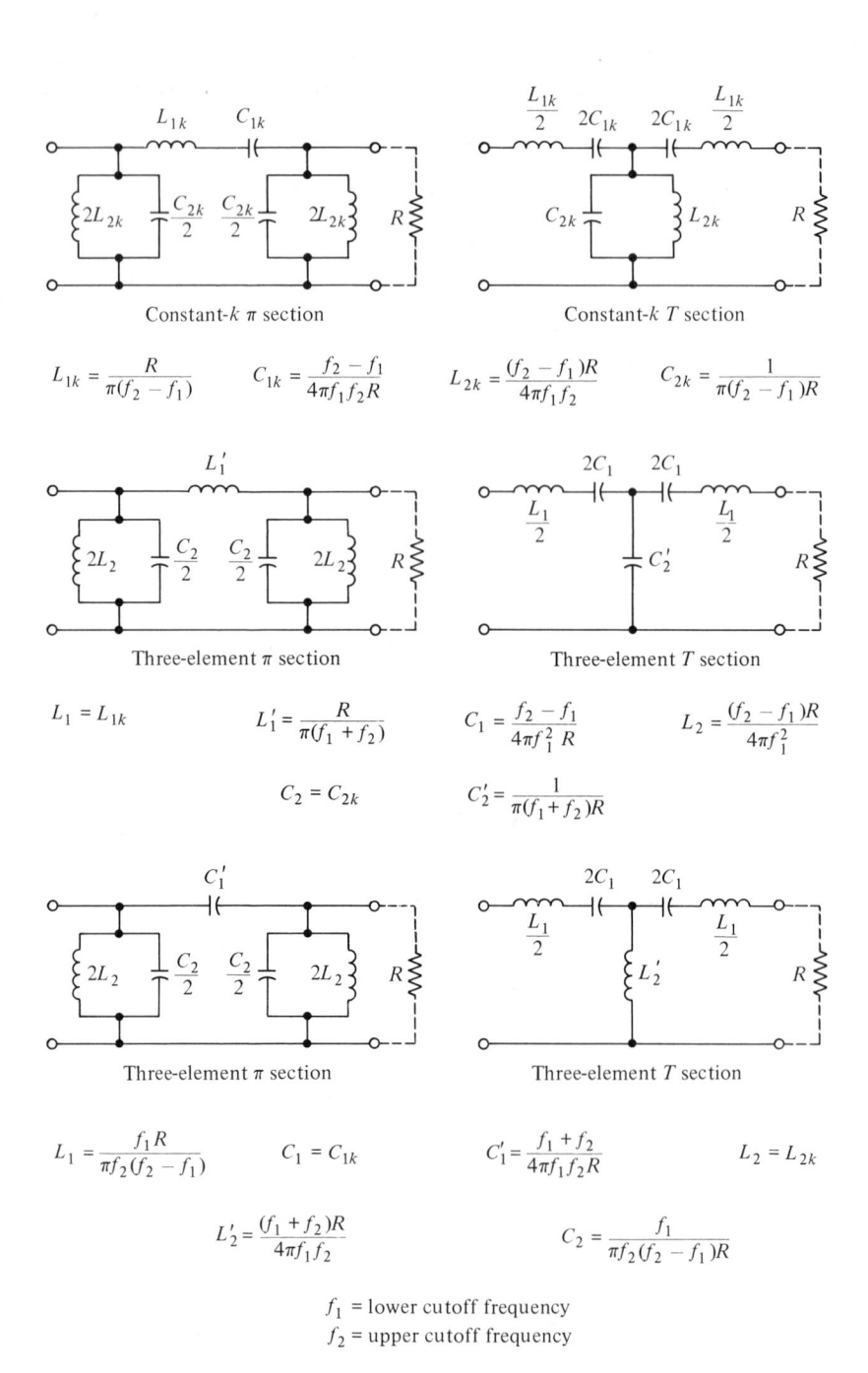

$$L_{2k} = \frac{(f_2 - f_1)R}{4\pi f_1 f_2} \qquad C_{2k} = \frac{1}{\pi(f_2 - f_1)R}$$

Three-element π section

$$L_1 = L_{1k} \qquad L_1' = \frac{R}{\pi(f_1 + f_2)} \qquad C_2 = C_{2k}$$

Three-element T section

$$C_1 = \frac{f_2 - f_1}{4\pi f_1^2 R} \qquad C_2' = \frac{1}{\pi(f_1 + f_2)R} \qquad L_2 = \frac{(f_2 - f_1)R}{4\pi f_1^2}$$

Three-element π section

$$L_1 = \frac{f_1 R}{\pi f_2 (f_2 - f_1)} \qquad C_1 = C_{1k}$$

$$L_2' = \frac{(f_1 + f_2)R}{4\pi f_1 f_2}$$

Three-element T section

$$C_1' = \frac{f_1 + f_2}{4\pi f_1 f_2 R} \qquad L_2 = L_{2k}$$

$$C_2 = \frac{f_1}{\pi f_2 (f_2 - f_1)R}$$

f_1 = lower cutoff frequency
f_2 = upper cutoff frequency

Figure 3–10 Configurations and design equations for basic bandpass filter sections.

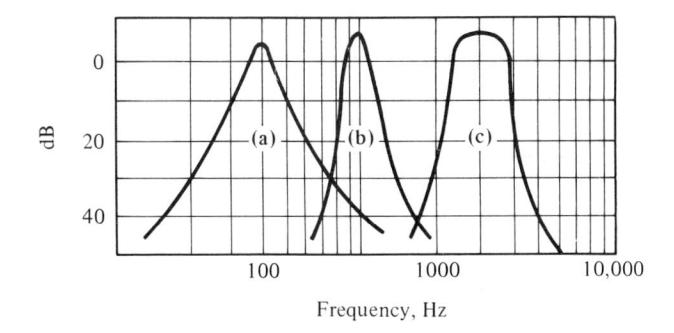

Figure 3-11 Frequency responses for single-section constant-k bandpass filters, and a two-section filter: **(a)** low-Q, narrow-band arrangement; **(b)** high-Q, narrow-band arrangement; **(c)** high-Q, wide-band, two-section arrangement.

is shown, with its m-derived form. In this example, m is assigned a value of 0.627 in the design equations for the m-derived bandpass filter configuration. These design equations describe an m-derived constant-k bandpass-filter section that incorporates the parameters of a prototype section supplemented by input and output m-derived end sections. The advantage of utilizing the m-derived design is seen in Fig. 3-12(c). In other words, the characteristic impedance through the pass band is more nearly constant; cutoff is much more abrupt, and the rate of rolloff is considerably greater.

It should be noted that the cutoff frequencies of the m-derived constant-k bandpass filter depicted in Fig. 3-12(b) are related to the value of m that is used in the design equations. The equation for m is written:

$$m = \sqrt{1 - \left[\frac{f_r (f_2 - f_1)}{f_r^2 - f_1 f_2} \right]^2}$$

where f_r is the resonant frequency of theoretically infinite attenuation.
f_1 is the lower cutoff frequency.
f_2 is the upper cutoff frequency.

In the design of a bandpass filter, the resonant frequency of infinite attenuation is usually taken as 2 percent beyond the upper cutoff frequency. In the example of Fig. 3-12, the upper cutoff frequency is 23 kHz. Thus, f_r will be assigned a value of 23.46 kHz.

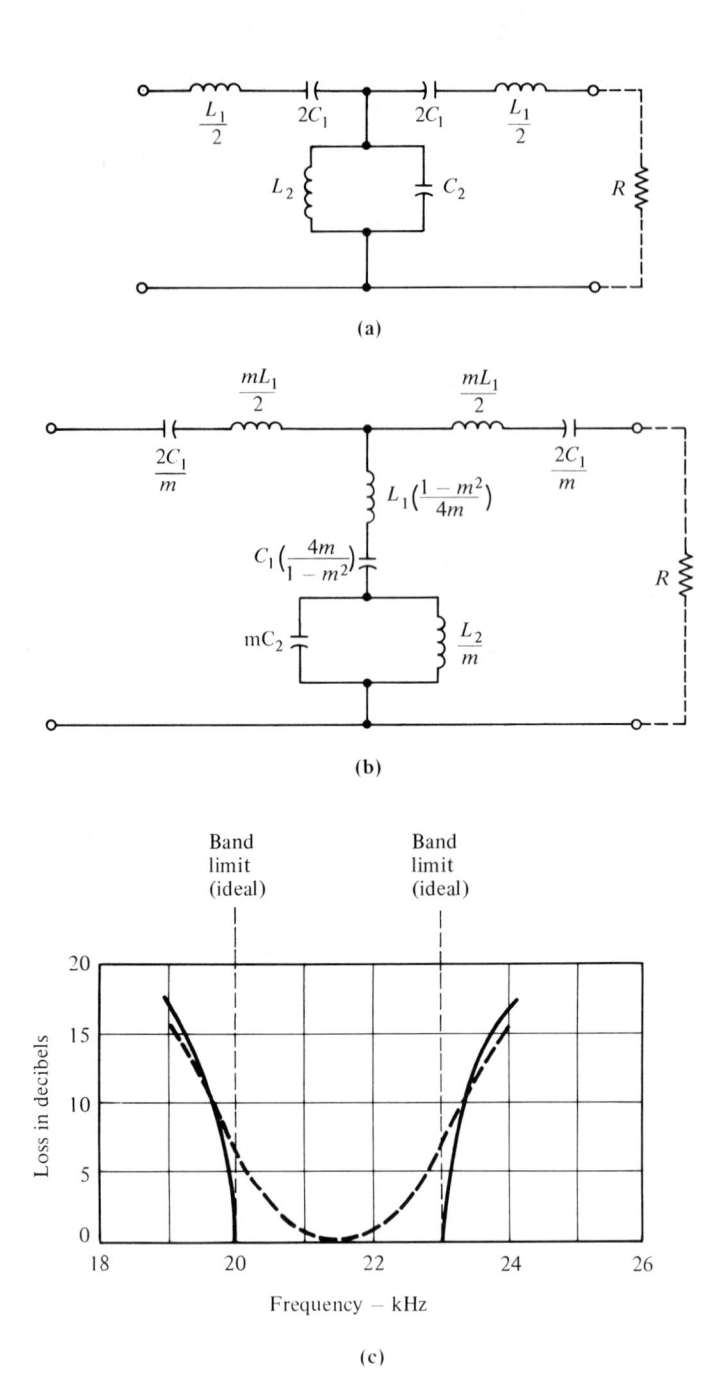

Figure 3-12 Comparison of prototype and *m*-derived constant-*k* bandpass filter configurations: **(a)** prototype circuit; **(b)** *m*-derived circuit; **(c)** comparative frequency responses.

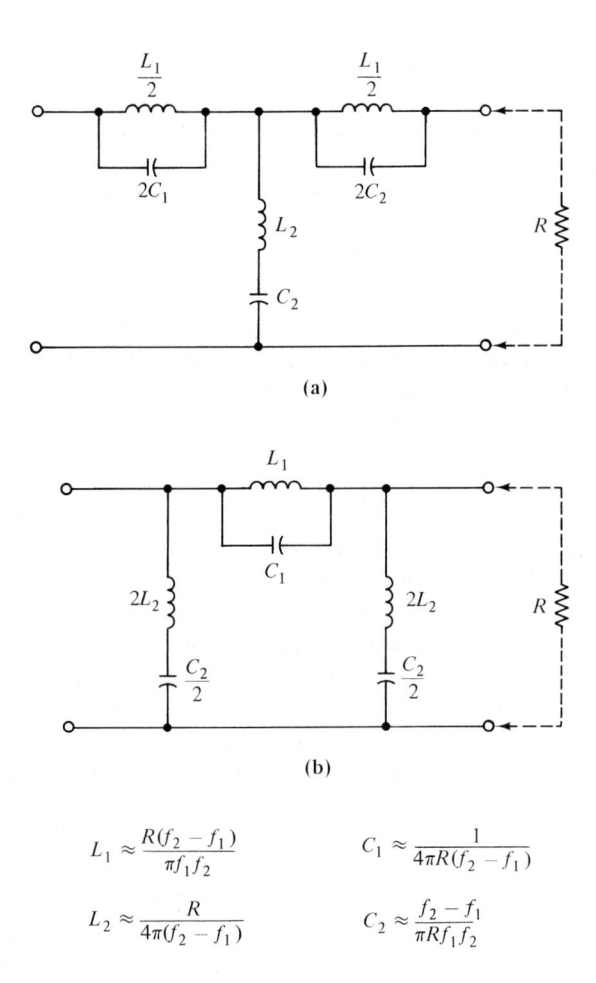

$$L_1 \approx \frac{R(f_2 - f_1)}{\pi f_1 f_2} \qquad C_1 \approx \frac{1}{4\pi R(f_2 - f_1)}$$

$$L_2 \approx \frac{R}{4\pi(f_2 - f_1)} \qquad C_2 \approx \frac{f_2 - f_1}{\pi R f_1 f_2}$$

Figure 3-13 Basic *LC* band-elimination wave-filter sections: **(a)** *T* section configuration; **(b)** pi section configuration.

Again, the lower cutoff frequency in this example is 20 kHz. When the foregoing equation is solved for *m*, the answer is 0.627. Note that this value of *m* also establishes another value for f_r or 19.608 kHz, which is approximately 2 percent below the lower cutoff frequency of 20 kHz. In summary, since a bandpass filter is assigned two cutoff frequencies, there are two associated frequencies of infinite attenuation. If we observe the configuration of the shunt arm in Fig. 3-12(b), it becomes evident that the circuit will exhibit two resonant frequencies.

Consider next the arrangement of a band-elimination wave filter. Basic constant-*k* pi and *T* sections are shown in Fig. 3-13; these may be compared with the bandpass configurations depicted in Fig. 3-10. From a qualitative viewpoint, the series and shunt arms have been interchanged in the two filter types. As would be anticipated, two band-elimination filters can be connected in cascade to obtain more rapid rolloff characteristics. Note that if a simple notch filter is formed by utilization of a single series-resonant circuit, or of a single parallel-resonant circuit, it is called a *wave trap,* or merely a *trap*. When a simple bandpass filter is formed by magnetically coupling a pair of parallel-resonant coils, it is termed a *tuned transformer*. Note in passing that although the secondary of a tuned transformer has the appearance of a parallel-resonant circuit, it operates as a series-resonant circuit by virtue of magnetic coupling action.

As noted previously, LC wave-filter calculations are based on the assumption that the inductors have negligible resistance. In many applications, this assumption is justified. The reactance of a coil with a given inductance value increases directly as the frequency increases. Its resistance remains essentially constant over the lower frequency range. On the other hand, at higher frequencies, the coil's AC resistance increases with an increase in frequency owing to *skin effect*. As a result, the Q value of an inductor will seldom exceed 400 in practice. One design approach to realize high Q values in wave filter operation is to use piezoelectric crystals such as quartz or barium titanate instead of conventional coils and capacitors. A quartz crystal may have an effective Q value in the order of 20,000 or more. In turn, bandpass filters with very steep rolloff characteristics can be structured with quartz or ceramic crystals. An alternative approach is the use of active devices in an otherwise conventional passive filter arrangement.

3-3 ELEMENTS OF ACTIVE FILTERS

An *active filter* exploits the characteristics of RC filter configurations in combination with one or more active devices such as transistors or operational amplifiers (op amps). One benefit of an active filter is the provision of comparatively steep rolloff characteristics. Another benefit is elimination of insertion loss. Some active filters are designed to provide substantial gain. For example, an active filter used as an audio preamplifier may have a maximum gain of 60 dB. A skeleton circuit for a low-pass active filter with a single op amp is shown in Fig. 3-14. At zero frequency (DC) the gain of the network is equal to:

Elements of Active Filters · 69

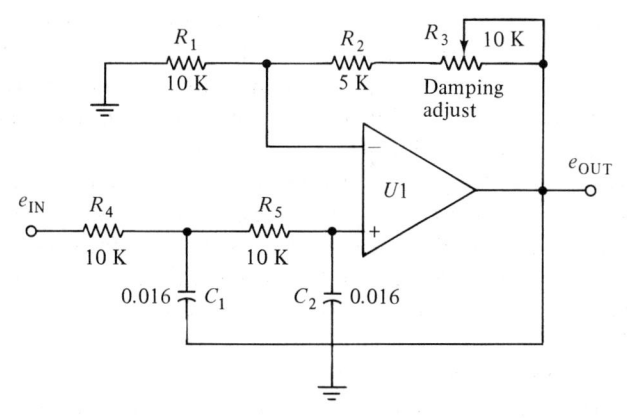

$$R_1 \quad 10\text{ K}$$
$$R_2 \quad 5\text{ K}$$
$$R_3 \quad 10\text{ K} \quad \text{Damping adjust}$$

e_{IN} R_4 10 K R_5 10 K U1 e_{OUT}

0.016 C_1 C_2 0.016

Lowpass, − 12dB/octave, 1 kHz

Figure 3–14 A low-pass active filter configuration.

$$A_V = \frac{e_{out}}{e_{in}} = \frac{R_1 + R_2 + R_3}{R_1} + 1$$

(RF series)

Thus, the zero-frequency gain is adjustable over a range from 2.5 to 3.5 times as the damping control R₃ is varied from its maximum to its minimum setting. This active filter has a rolloff rate of 12 dB per octave; this is a much more rapid rolloff than can be obtained with passive circuitry of the same order of complexity. The rapid rolloff is the result of the positive-feedback loop from the output of U₁ through C₁, through R₅, over C₂, and into the noninverting input terminal of U₁. In the first analysis, this positive feedback overcomes the insertion loss of the passive filter sections. It also compensates for the tendency of a passive filter to exhibit a droop of the frequency-response curve through the pass band. This compensation results from decreasing reactance of C₁ as the frequency increases, with an accompanying increase in positive feedback.

Note that C₁ has a reactance of approximately 10,000 ohms at the filter cutoff frequency of 1 kHz. At still higher frequencies, the reactance of C₁ is less, and a larger fraction of the output from U1 is fed back. Accordingly, it is evident that the low-pass action of R₅ and C₂ is opposed by the positive-feedback action of C₁. However, as the frequency increases, the two-section low-pass action attenuates the input to U1 at a greater rate than the feedback from the output of U1 increases. In consequence, the output from U1 rolls off at an increasing rate as the frequency increases. As noted previously, the filter output rolls off 12 dB per octave. The negative-feedback level is adjusted by means of the damping control R_3. This negative-feedback loop has no frequency discrimination. In effect, the amount of positive feedback varies inversely with the negative-feedback level that is utilized. If R_3 is set for minimum negative feedback, the filter frequency response rises to a peak as the cutoff region is approached. On the other hand, if R_3 is set for maximum negative feedback, the filter frequency response droops as the cutoff region is approached. At a suitable intermediate setting of R_3, the frequency response remains flat from zero frequency up to the start of rolloff.

Next, consider the high-pass active filter configuration shown in Fig. 3–15. This arrangement is similar to that of the low-pass active filter discussed above, with the capacitors and resistors of the RC sections interchanged. The rolloff rate of the high-pass arrangement is 12 dB per octave, and its cutoff frequency is 1 kHz. Positive feedback from the output of U1 through R_4 and finally to the noninverting input of the op amp acts to produce the comparatively rapid rate of rolloff. Because of the gain provided by U1, there is no insertion loss from input to output of the filter, within its pass band. Positive feedback also compensates for the tendency of the passive filter sections to exhibit a droop of the frequency-response curve through the pass band. Maximum flatness of response is obtained by suitable adjustment of the damping control R_3.

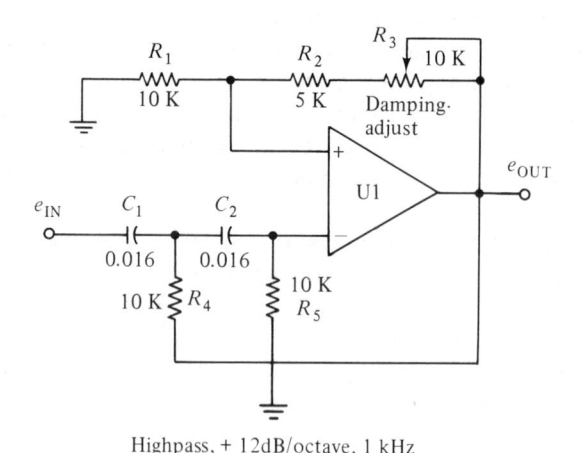

Highpass, + 12dB/octave, 1 kHz

Figure 3–15 A high-pass active filter configuration.

The cutoff frequency of the configuration in Figs. 3-14 and 3-15 is determined by the capacitance values in the *RC* filter sections. To reduce the cutoff frequency of either configuration, use higher values of capacitance. Or, to increase the filter cutoff frequency, use lower values of capacitance. Because these active-filter arrangements use a very large amount of negative feedback, they are quite stable. The filter characteristics depend essentially upon the values of the resistors and capacitors, and not upon the gain of the op amp. Of course, if the op amp deteriorates seriously, so that its gain is greatly decreased from bogie value, the characteristics of the active filter will begin to be affected. In design of the circuit, no account is taken of op-amp gain value. Op amps used in this general class of active filters are called *operational voltage amplifiers* (OVA's), and they have an extremely high value of input impedance. Their output impedance is low; a value of 150 ohms is typical.

Consider next the simple active bandpass-filter configuration shown in Fig. 3-16. A bandpass characteristic is obtained by means of an *RC* filter network in the negative-feedback loop for the op amp. At very low frequencies, C_1 is effectively an open circuit, and no signal is applied to the inverting input of the op amp. As the input signal frequency is increased, the reactance of C_1 decreases, and a small input signal is applied to U_1. This input signal is amplified and appears at the output of U_1. Part of the output signal is fed back via C_2, but there is comparatively little negative-feedback action at this frequency, owing to the high reactance of C_2 compared with the internal resistance of the source. Thus, the output from U_1 proceeds to increase rapidly as the signal frequency increases. However, there

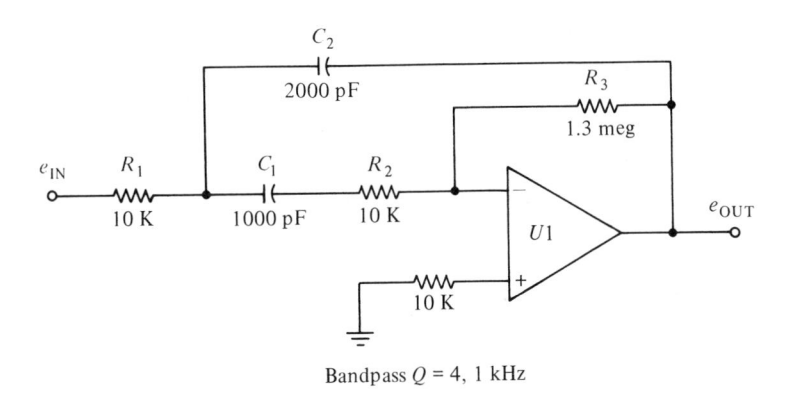

Bandpass $Q = 4$, 1 kHz

Figure 3-16 A simple active bandpass-filter configuration.

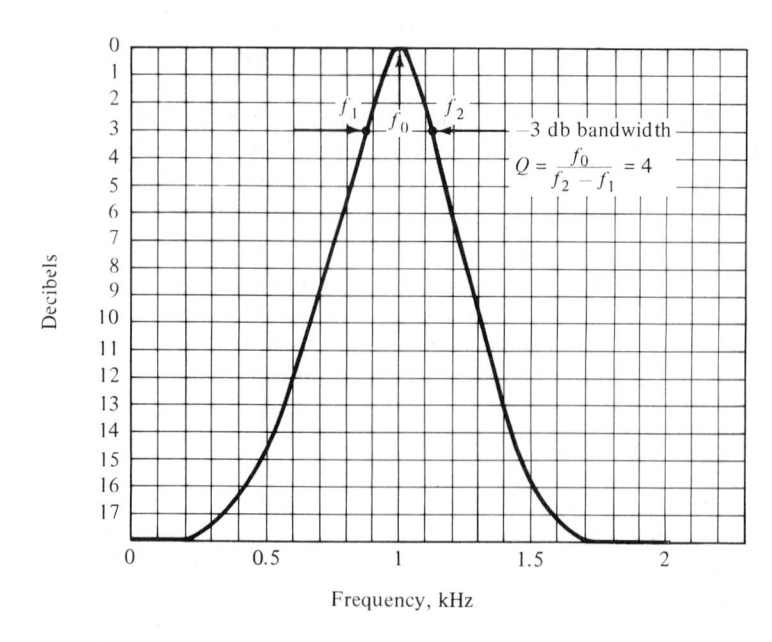

Figure 3-17 Frequency-response curve with a Q value of 4.

Graph shows frequency response with peak at f_0, with f_1 and f_2 marking the -3 dB bandwidth points, and $Q = \dfrac{f_0}{f_2 - f_1} = 4$.

is a limiting frequency (1 kHz) at which negative-feedback action inhibits any further increase in the output signal level.

At 1 kHz, C_2 has a reactance of approximately 80,000 ohms. In turn, the increased negative feedback prevents any increase in output from U1. At any frequency greater than 1 kHz, the increasing negative feedback progressively decreases the output from U_1, and the op-amp gain approaches unity as a limit. This frequency-dependent negative-feedback action results in a frequency-response curve such as that depicted in Fig. 3-17. The −3-dB bandwidth is 250 Hz, which corresponds to a system Q value of 4. That is, the Q value is approximately equal to the quotient of the peak frequency and the −3-dB bandwidth. Note that this frequency response could also be obtained with a single-stage tuned amplifier. However, the active-filter arrangement has an advantage, in that it avoids the necessity for a large inductor. Just as passive filters can be cascaded to obtain steeper rolloffs or narrower bandwidths, so may active filters be cascaded.

Active filters have an advantage over passive filters in cascade arrangements, because the former offer the possibility of interactive configurations that provide operational flexibility. For example, a cascaded interactive bandpass active filter is exemplified in Fig. 3-18, with three op amps supplemented by comparatively simple *RC* circuitry. A potentiometer is included in the negative-feedback network

Elements of Active Filters · 73

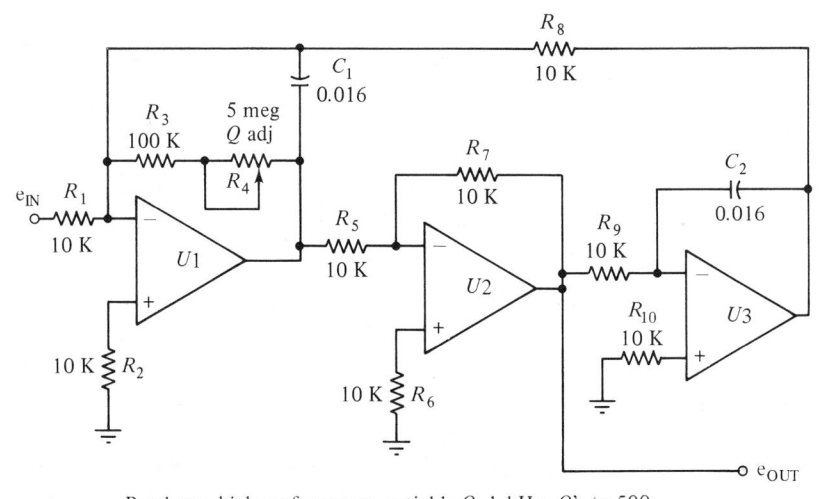

Bandpass, high performance, variable Q, 1 kHz, Q's to 500

Figure 3–18 A three-stage active bandpass-filter configuration.

for U1; this is a Q-adjustment control which permits the bandpass response to be varied from a comparatively large bandwidth to a very narrow bandwidth. Thus, a maximum Q value of 500 is available; this corresponds to a −3-dB bandwidth of approximately 2 Hz at a peak frequency of 1 kHz. As in the simpler arrangement of Fig. 3–16, the three-stage active filter does not utilize positive feedback—only negative feedback is used. Note that the first and third stages operate with frequency-dependent feedback; on the other hand, the second stage operates as a frequency-independent buffer.

Component and Device Tolerances Tolerances on op-amp gain can be neglected in most applications; utilization of a large amount of negative feedback ordinarily makes the stage gain a function of input-component and feedback-component values from a practical viewpoint. For example, consider the basic op-amp amplifier arrangement depicted in Fig. 3–19. In this example, U1 has a maximum available gain (MAG) of 100,000 times. However, with the negative-feedback resistor R_f included in the circuit, the stage amplification with feedback, A_{fb}, becomes essentially equal to

$$A_{fb} = \frac{R_i + R_f}{R_i}$$

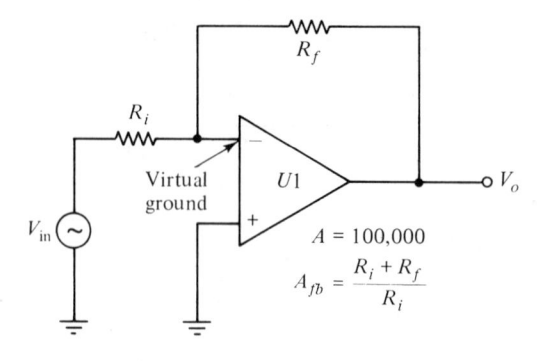

```
                    Rf
                   ─WW─
           Ri
          ─WW─●
              Virtual    U1
              ground
      Vin ~        +              ──o Vo
                         A = 100,000
                         Afb = (Ri + Rf)/Ri
```

Figure 3-19   Basic op-amp amplifier arrangement.

Thus, if R_f and R_i each have a value of 10,000 ohms, the stage gain will be equal to two times, from a practical viewpoint. If U1 were replaced with another op amp that had a maximum available gain of 150,000 times, the stage gain would still be equal to two times from a practical viewpoint. This fact becomes evident from an inspection of the general equation for the gain of a stage with negative feedback:

$$A_{fb} = \frac{A}{1 + A\frac{R_i}{R_f + R_i}}$$

Observe that if A is equal to 100,000, R_i is 10,000 ohms, and R_{fb} is 10,000 ohms, A_{fb} becomes 1.9999 ... out to many decimal places. Then, if A is made equal to 150,000, A_{fb} again becomes 1.9999 ... out to many decimal places. Therefore, when the circuit designer utilizes op amps in active filters, *tolerances on op-amp gain can ordinarily be neglected.*

Next, consider the input resistance to an op amp with a large amount of negative feedback, and the tolerance on this input resistance. With reference to Fig. 3-19, U1 presents a *virtual ground* at its inverting input, owing to substantial negative feedback. In turn, the input resistance to the op amp becomes equal to R_i for all practical purposes. In turn, the tolerance on the input resistance to the op amp is the same as the tolerance on R_i. On the other hand, there is a substantial tolerance on the input impedance of the noninverting input, because negative feedback is not applied to this input terminal. (Positive feedback may be applied to the noninverting input in some active-filter arrangements.) Thus, the input impedance at the noninverting input may range from 150,000 ohms to 1.5 megohms for a typical op amp.

The value of A_{fb} can be appreciably affected by tolerances on R_f and R_i. For example, suppose that both resistors have a tolerance of ±20 percent, and equal bogie values. In turn, the *worst-case condition* occurs when R_i has its lower tolerance limit and R_f has its upper tolerance limit. In this case, the value of A_{fb} becomes 25 percent higher than bogie. As another illustration, if R_f has twice the value of R_i, and both resistors have ±20 percent tolerances, the worst-case condition occurs when R_i has its lower tolerance limit and R_f has its upper tolerance limit. In this example, the value of A_{fb} becomes 33 percent higher than bogie.

Op-amp Design Guidelines Op amps can be damaged by improper design of active-filter circuitry. An op amp is rated for a minimum value and a maximum value of power-supply voltage. For example, the maximum rated supply voltage is +10 volts, and the recommended minimum supply voltage is +5.5 volts, for a typical op amp. The device will not operate normally if supplied with less than +5.5 volts. On the other hand, if supplied with more than +10 volts, the collector-base junctions of the integrated-circuit transistors will break down, and the op amp will be destroyed by excessive current flow.

Damage can also result from excessive drive signal voltage. The base-emitter junctions of the input stage in the op amp are likely to break down if the maximum rated input voltage is exceeded. When a large value of capacitance is employed in the negative-feedback loop of an op amp, power-supply transients may present a pitfall that causes destruction of the op amp when the circuit is turned on or off. In other words, a large charge storage in the feedback capacitor may drive the input of the op amp past its maximum rated input voltage under this condition. This danger is not present with all types of op amps; however, *if the maximum rated input voltage is less than the power-supply voltage, the designer should be on guard for this pitfall.*

Most op amps will be seriously damaged if the polarity of the power-supply voltage is reversed, even for an instant. This catastrophe is the result of excessive current flow through the isolation junctions in the op-amp structure, which become forward-biased if the power-supply voltage is reversed in polarity. Therefore, it is good foresight to provide protective diodes in the power-supply line(s) if there is any possibility of accidental power-supply polarity reversal in operation of the active filter. This danger is most likely to be encountered in the design of battery-operated equipment.

An op amp is rated for maximum total power dissipation, such as 300 mW. Most op amps are fabricated in a manner that prevents the maximum power rating from being exceeded, provided that the power-supply voltage(s) and the input signal voltage(s) are within their maximum ratings. However, there are a few types of op amps that are subject to damage if operated at maximum ratings with the output terminal short-circuited to ground. Therefore, if this hazard should exist under any condition of operation in an active filter, the circuit designer should include a protective series resistor in the output lead from the op amp. This resistor should have a value sufficient to prevent excessive device dissipation under the condition of a short-circuited output.

Most op amps are immune to *latch-up*. However, this possibility of active-filter malfunction may be encountered with a few types of op amps. When latch-up occurs as a result of exceeding the maximum rated input signal voltage, the output circuitry in the op amp becomes "stuck," or "locked up." In turn, the output terminal rests at either the positive or the negative maximum output voltage, and the op amp is thereafter unresponsive to any change of input voltage. If this hazard exists in the design of an active filter, it is advisable to protect the op-amp input(s) with zener diodes to limit the maximum input voltage that can be applied.

3–4 INCOMING INSPECTION PROCEDURES

Under various conditions, the circuit designer or the chief engineer may elect to establish incoming inspection procedures for components, devices, or subassemblies. For example, it may be less costly to purchase a shipment of wide-tolerance components and to screen them through incoming inspection, than to pay a premium for close-tolerance components. Again, a production change order may have been issued for tighter-tolerance components in a production run. To avoid assembly-line downtime while awaiting a new shipment, an incoming inspection procedure may be stipulated. In any case, a test procedure is devised that can be conducted reliably by comparatively unskilled personnel. Thus, a go/no-go type of test is utilized, if possible. Fixed control settings are specified for generators, meters, oscilloscopes, and power supplies. The component or device under test is plugged into a jig or receptacle, and the result of the test is observed on a meter scale or on an oscilloscope screen. Accept-reject limits are marked on the meter window or the oscilloscope screen, so that no interpretation of test results is required of the operator.

CHAPTER 4

Basic Audio Amplifier Design

4–1 GENERAL CONSIDERATIONS

A typical audio-amplifier channel in a high-fidelity system has a maximum usable gain (MUG) of more than 5000 times, as depicted in Fig. 4–1. Although no industry standards have been established, it is generally considered that high-fidelity reproduction entails a frequency response from at least 20 Hz to 20 kHz, with a total harmonic distortion of less than 1 percent at maximum rated power output. The basic distortion measurement is made at a frequency of 1 kHz. *Power bandwidth* is defined as the frequency range between an upper limit and a lower limit, at a power level 3 dB below maximum rated power output, where the harmonic distortion starts to exceed the value that occurs at a frequency of 1 kHz and at maximum rated power output. The *music-power rating* of an amplifier is defined as the peak power that can be delivered to the speaker for a very short period of time, with no more harmonic distortion than at the maximum rated sine-wave rms output. In other words, a music-power rating denotes the ability of an amplifier to process sudden peak musical waveforms without objectionable distortion. The peak duration in this mode of operation is limited by the ability of the filter capacitors in the power supply to sustain the peak current demand.

Most high-fidelity amplifiers can be classified into preamplifier and power-amplifier types. Typical input/output characteristics for a preamplifier are noted in Fig. 4–2. Four inputs are provided, to ac-

78 · Basic Audio Amplifier Design

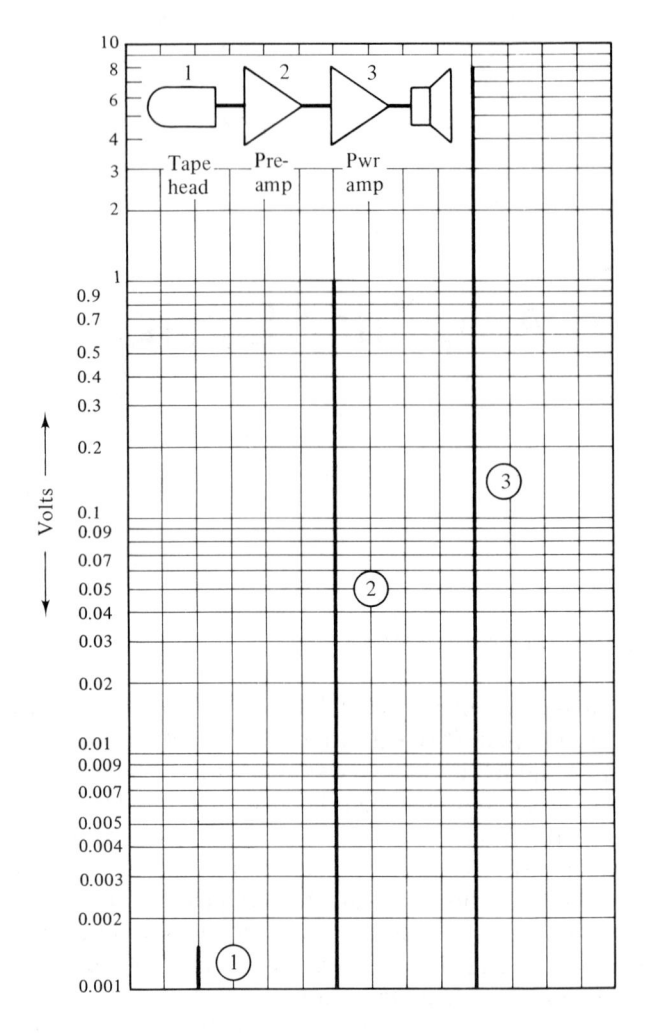

Figure 4–1 Signal-voltage levels in a typical high-fidelity channel.

commodate a low-impedance microphone, a tape player, a phono player, and an FM tuner. Each input port is rated for a different signal level. Thus, the rated input signal level for the low-impedance microphone port is 350 μV, whereas the rated level for the FM tuner port is 250 mV, or a range of 700 to 1. Note also that the frequency response of the amplifier is different for each input port. The maximum available voltage gain of the exemplified preamplifier is 69 dB, and an output of 1 V rms is normally delivered to a 10,000-ohm

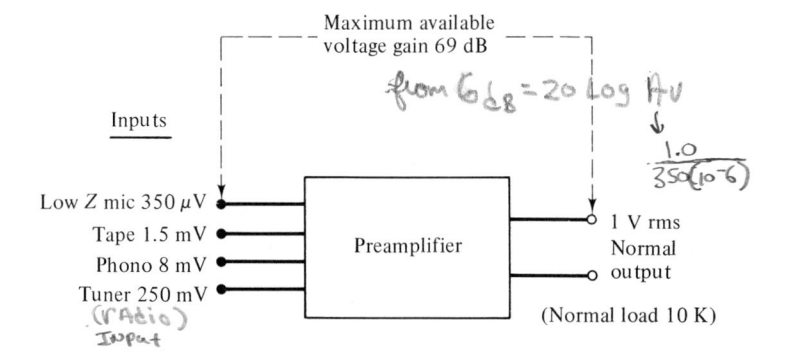

```
                    Maximum available
              ┌─────voltage gain 69 dB─────┐
              │      from 6 dB = 20 log A_v │
   Inputs     │                             │
              │                            1.0
                                          ─────────
                                          350(10⁻⁶)
  Low Z mic 350 μV ●─┐
  Tape 1.5 mV      ●─┤                    ○ 1 V rms
  Phono 8 mV       ●─┤   Preamplifier        Normal
  Tuner 250 mV     ●─┤                    ○ output
   (radio)          └──────────────────┘  (Normal load 10 K)
   Input
```

Figure 4–2 Typical voltage gain values for a hi-fi preamplifier.

load, regardless of which input port is in use. Next, consider the typical voltage-gain and power-output values for a power amplifier, as noted in Fig. 4–3. A signal-voltage gain of 18 dB is provided, and an input of 1 V rms produces approximately 8 V rms across an 8- or 16-ohm load. Since the input impedance of the amplifier is 10,000 ohms, its power gain is considerable, although its voltage gain is only moderate.

4–2 BASIC AMPLIFIER CHARACTERISTICS

It is instructive to observe the basic amplifier characteristics for small-signal bipolar transistors shown in Fig. 4–4. The three basic configurations are characterized for a load resistance of 10,000 ohms and a generator internal resistance of 1000 ohms. Thus, the common-base (CB) configuration has the highest voltage gain, the common-collector (CC) configuration has the highest current gain, the common-emitter (CE) configuration has the highest power gain, the CC configuration has the highest input resistance, and the CB configuration has the highest output resistance. Note that the input-resistance and output-resistance values denote AC resistance—not DC resistance. An AC resistance is also called a *dynamic resistance* or an *incremental resistance*. Since the CE configuration has the least difference between its input- and output-resistance values, it is the most commonly utilized amplifier configuration.

Observe that the power gain of the stage is equal to the product of its voltage gain and current gain. Although the CB configuration has a

80 · Basic Audio Amplifier Design

```
            Voltage gain
      ┌ ─ ─  18 dB  ─ ─ ┐
      │                  │
Input │                  │
──────┤                  │
      ▼                  ▼
    ┌─────────────────────┐
    │                     │──o 8 V rms (approx.)
1 V rms│    Power         │
    │    amplifier        │   5 watts
    │                     │──o normal
    │                     │    output
    └─────────────────────┘
    (Input resistance      (Normal load
     = 10 K, approx.)       8 or 16 ohms)
```

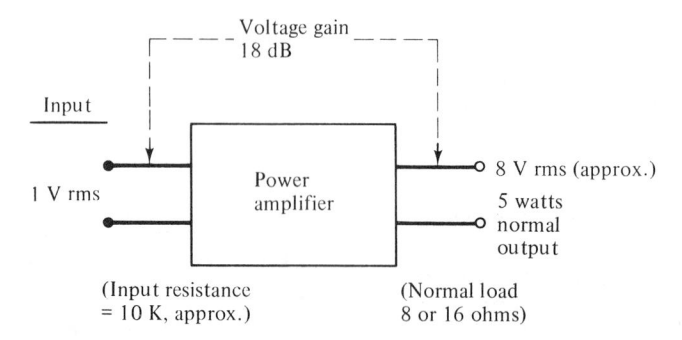

Figure 4–3 Typical voltage-gain and power-output values for a hi-fi audio power amplifier.

very large difference between its input- and output-resistance values, this relation is sometimes useful for impedance matching in circuit designs. Since the CC configuration has the lowest output resistance, it is useful as an impedance transformer when a high-impedance source must supply substantial signal power to a comparatively low-resistance load. As shown in Fig. 4–5, the power gain in a particular configuration varies considerably with changes in load resistance over a range from 10 ohms to 1 megohm. The 10,000-ohm value on which the data in Fig. 4–4 are based is a typical bogie value for the output resistance of a preamplifier and for the input resistance of a power amplifier. Note that the maximum power gain (MPG) for the CE configuration occurs for a load resistance of approximately 50,000 ohms. (An input level from 1 μV to 10 mV is termed small-signal operation.)

Component and Device Tolerances It is helpful to consider the effect of component and device *tolerances* on signal power output for the CE configuration. With reference to Fig. 4–5, it is seen that a ±20 percent tolerance on a 10,000-ohm load resistor will not result in a large tolerance on signal-power output. On the other hand, the tolerance on the current-gain (beta) value of the transistor results in a large tolerance on signal-power output. As noted above, the power gain for an amplifier stage is equal to the product of its voltage gain and current gain. For example, the power gain is proportional to the square of current, and is proportional to the square of voltage. Thus, a doubling of the current gain, accompanied by a doubling of the voltage gain in a stage, results in a quadrupling of the power gain. Suppose that the transistor in the stage has a beta value of 52.5 ±17.5 percent. In turn, its beta value may range from 35 to 70. If it

Basic Amplifier Characteristics · 81

(a) Common emitter

Voltage gain: 270 times
Current gain: 35 times
Power gain: 40 dB
Input resistance: 1.3 K
Output resistance: 50 K
(For generator internal resistance of 1 K)

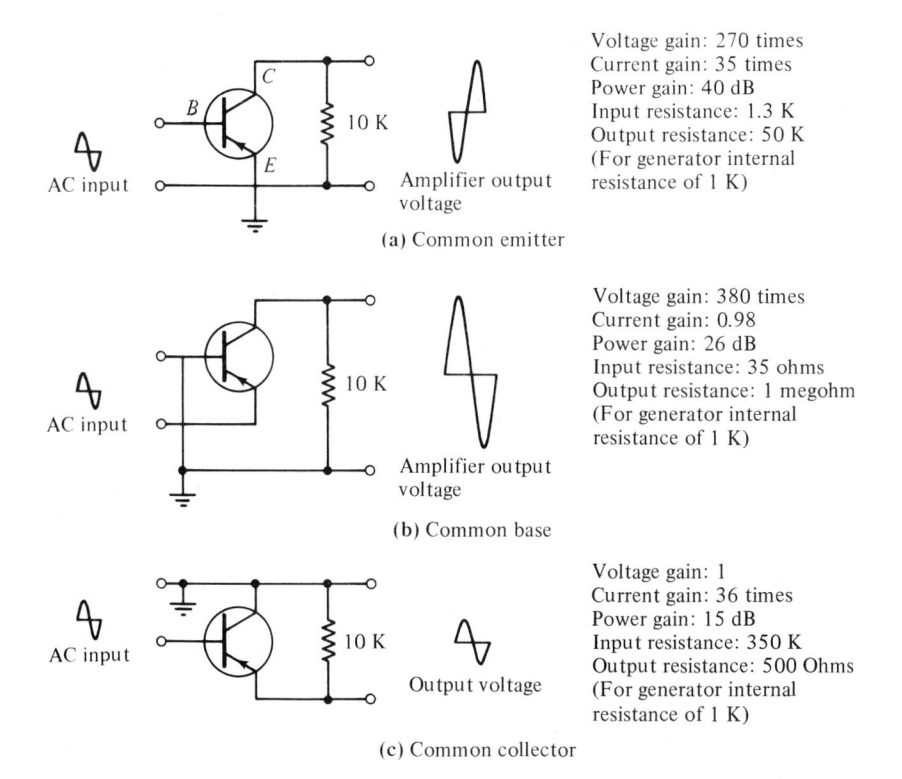

(b) Common base

Voltage gain: 380 times
Current gain: 0.98
Power gain: 26 dB
Input resistance: 35 ohms
Output resistance: 1 megohm
(For generator internal resistance of 1 K)

(c) Common collector

Voltage gain: 1
Current gain: 36 times
Power gain: 15 dB
Input resistance: 350 K
Output resistance: 500 Ohms
(For generator internal resistance of 1 K)

Figure 4–4   Basic amplifier characteristics for small-signal bipolar transistors.

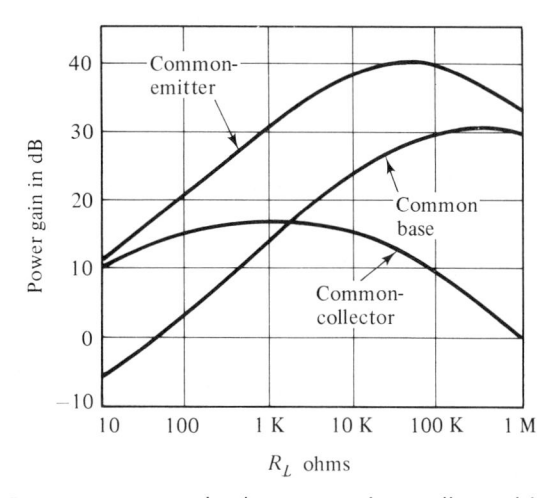

Figure 4–5   Power gain versus load resistance for small-signal bipolar transistors.

is assumed that stage operation is linear, a doubling in device current gain is accompanied by a doubling of stage current gain, accompanied by a doubling of stage voltage gain. In turn, the power gain of the stage is quadrupled, or a tolerance of ±17.5 percent on the transistor results in a tolerance of ±60 percent on the stage power output.

4–3 INPUT RESISTANCE VERSUS LOAD RESISTANCE

As shown in Fig. 4–6, the input resistance of a CE stage is almost independent of the load-resistance value. On the other hand, the input resistance of a CB stage or of a CC stage will vary greatly from small to large values of load resistance. Note that the horizontal axis is basically numbered in R_L/r_c units, where R_L is the load-resistance value, and r_c is the "collector resistance" of the transistor. This latter value is very large, in the order of 1 or more megohms. If it is assumed that the transistor has a beta value of 50, the horizontal axis may be alternatively numbered in R_L units, as exemplified in Fig. 4–6. Observe that the input resistance of various transistors is a function of frequency. For example, a typical germanium PNP alloy drift-field transistor that has an input resistance of 1350 ohms at low frequencies and up to 1.5 MHz exhibits an input resistance of only 150 ohms at 12.5 MHz.

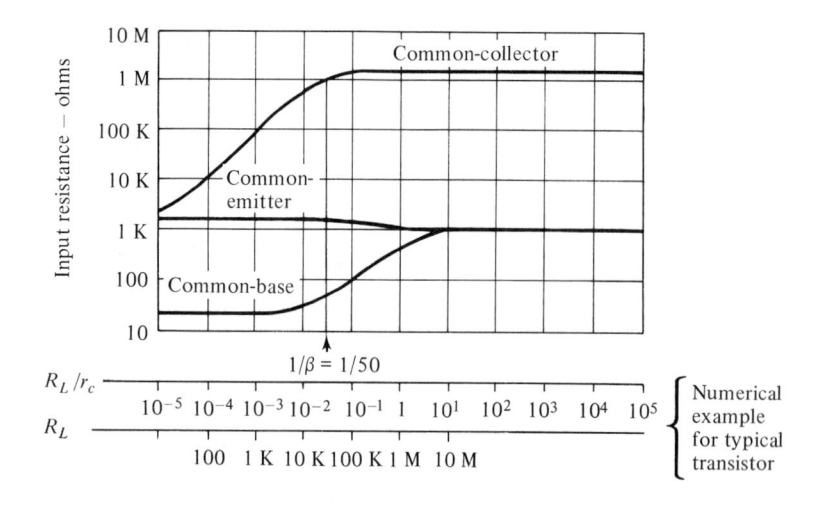

Figure 4–6 Input resistance versus load resistance for a small-signal bipolar transistor.

Voltage and Current Amplification · 83

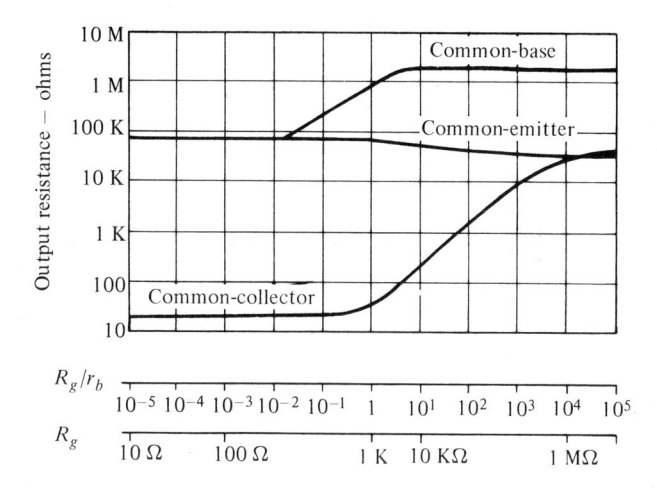

Figure 4-7 Output resistance versus generator resistance for a small-signal bipolar transistor.

4-4 OUTPUT RESISTANCE VERSUS GENERATOR RESISTANCE

As shown in Fig. 4-7, the output resistance of a CE stage is fairly independent of the generator-resistance value. On the other hand, the output resistance of a CB stage or of a CC stage will vary greatly from small to large values of generator resistance. Note that the horizontal axis is basically numbered in R_g/r_b units, where R_g is the generator-resistance value, and r_b is the base resistance of the transistor. If it is assumed that the transistor has "average" characteristics, the horizontal axis may be alternatively numbered in R_g units, as exemplified in Fig. 4-7. The output resistance of various transistors is also a function of frequency. As an example, a transistor that has an output resistance of 70,000 ohms at low frequencies and up to 1.5 MHz, exhibits an output resistance of only 4000 ohms at 12.5 MHz.

4-5 VOLTAGE AND CURRENT AMPLIFICATION

At high values of load resistance, the voltage amplification of a stage operating in the CE, CB, or CC mode attains a plateau, as seen in Fig. 4-8. However, at lower values of load resistance, the voltage amplification decreases. The CE and the CB configurations have similar trends in voltage amplification, although the gain of a CB

84 · Basic Audio Amplifier Design

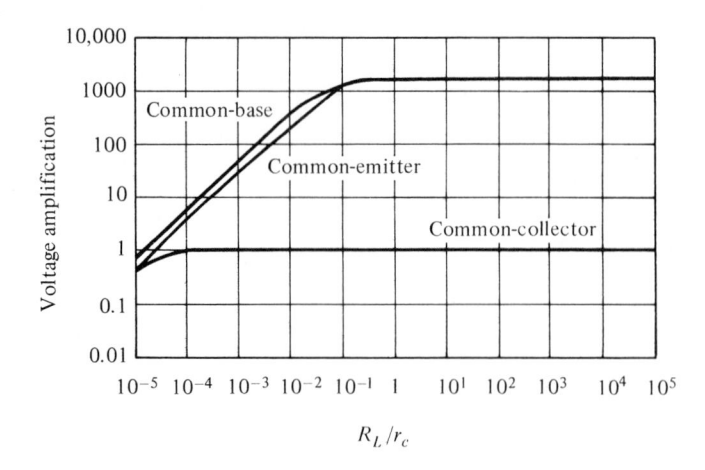

Figure 4–8 Voltage amplification versus load resistance for a small-signal bipolar transistor.

stage is slightly higher than that of a CE stage for lower values of load resistance. As would be anticipated, the CC configuration has a voltage amplification of practically unity for all except very low values of load resistance. At very low values, the load approaches a short-circuit condition, and the voltage amplification trends to zero. Note in passing that uniformity of voltage amplification does *not* correspond to uniformity of power amplification in any of the three operating modes. This lack of correspondence results from the fact that the current amplification of a stage varies most over the load-resistance range where the voltage amplification varies least, as will be explained next.

Variation of current amplification versus load resistance for the CE, CB, and CC modes is shown in Fig. 4–9. In contrast to the voltage-amplification trend, the current amplification of a stage attains a plateau at low values of load resistance. Since the power amplification of a stage is equal to the product of its voltage amplification and its current amplification, the power output attains a maximum value when a suitable value of load resistance is utilized, as was shown in Fig. 4–5. Different values of load resistance are required to obtain maximum power amplification in the CE, CB, and CC modes of operation. In many circuit-design projects, the chief requirement in stage operation is to obtain maximum power output.

Basic FET Amplifier Characteristics · 85

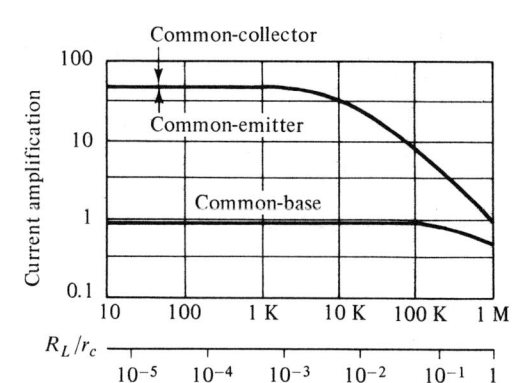

Figure 4–9 Current amplification versus load resistance for a small-signal bipolar transistor.

Both the voltage amplification and the current amplification of a stage will vary in accordance with the *tolerance on the beta value* of the transistor. On the other hand, tolerances on the load-resistance value have much less effect on voltage and current amplification. Transistors in most production lots have a comparatively wide tolerance on beta value. Accordingly, the circuit designer ordinarily chooses somewhat elaborate amplifier circuitry that tends to minimize the effect of lack of uniformity in beta values. However, design situations are encountered in which a pair of transistors must have beta values that are practically the same, if the stage is to perform satisfactorily. Such selected transistors are termed *matched pairs*.

4–6 BASIC FET AMPLIFIER CHARACTERISTICS

Field-effect transistors (FET's) are basically different from bipolar transistors in that the gate is isolated in the device structure and does not draw current. Three basic FET amplifier configurations are employed, as shown in Fig. 4–10. The common-source mode is in widest use. Since the gate of the FET draws negligible current, it is a *voltage-operated device,* whereas a bipolar transistor is a *current-operated device.* These terms are only relative, but they indicate the basic distinction between the two types of devices. Thus, the gain of an FET is stated as a *transconductance* value, whereas the gain of a bipolar transistor is stated as a *beta* value. The FET depicted in Fig.

86 · Basic Audio Amplifier Design

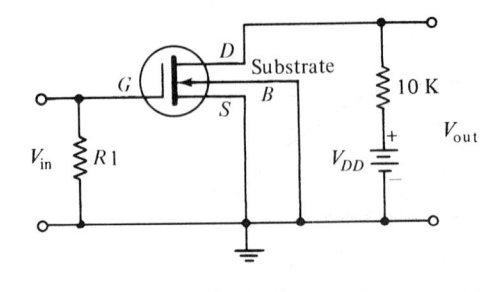

(a) Common source

Voltage gain: 50 times
Transconductance: 5000 μmhos
Power gain: 17 dB (50 times)
Input resistance: very high
Output resistance: 20 K
(For generator internal resistance of 500 ohms)

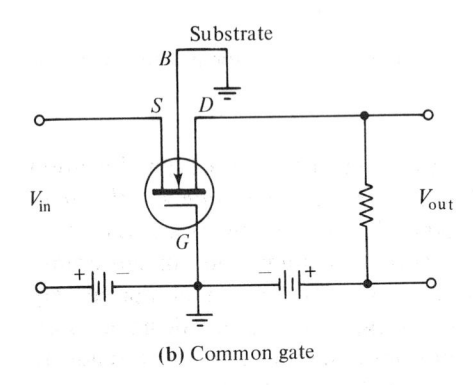

(b) Common gate

Voltage gain: 1.8
Input resistance: 240 ohms
Output resistance: high
(For generator internal resistance of 500 ohms)

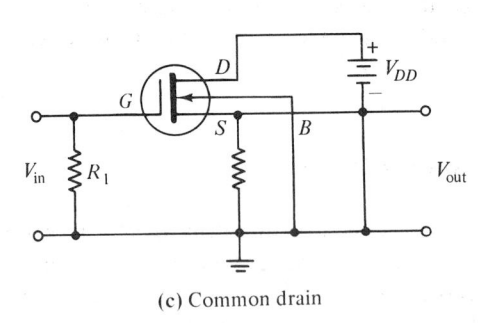

(c) Common drain

Voltage gain: 0.5
Input resistance: 2 M
Output resistance: 240 ohms
(For generator input resistance of 500 ohms)

Figure 4–10 Basic FET amplifier configurations and characteristics.

4–10 has a transconductance (Gm) value of 5000 μmhos. In turn, if the stage has a 10,000-ohm load resistor, the stage voltage gain will be 50 times. Note that the power gain is 17 dB, which is considerably less than the 40 dB provided by a corresponding CE bipolar transistor amplifier stage. (See also Chart 4–1.)

CHART 4-1: Comparative Device Parameters

Parameter	Bipolar Transistor	JFET Transistor	MOSFET Transistor
Input Impedance	Low	High	Very High
Noise	Low	Low	Unpredictable
Aging	Not Noticeable	Not Noticeable	Noticeable
Bias Voltage Temperature Coefficient	Low and Predictable	Low and Predictable	High and Unpredictable
Control Electrode Current	High	0.1 nA	10 pA
Overload Capability	Comparatively Good	Comparatively Good	Poor

However, the circuit designer will often choose an FET stage instead of a bipolar transistor stage when very high input resistance is required. The output resistance of an FET stage is substantially less than that of a corresponding CE bipolar transistor amplifier stage. This is ordinarily an unimportant consideration. Note that there are two basic subclasses of FET's. The *depletion* type of FET draws substantial current from V_{DD} when there is no input signal present. On the other hand, the *enhancement* type of FET draws practically no current from V_{DD} when no input signal is applied. Audio amplifiers generally are designed with depletion-type FET's. However, digital-logic circuitry is ordinarily designed with enhancement-type FET's. The chief advantage of enhancement FET's in logic systems is higher operating efficiency.

4-7 LOAD LINES

Transistors are inherently nonlinear devices. However, the designer is justified in assuming that a transistor is a linear device in small-signal operation. This assumption, on the other hand, would lead to gross errors in large-signal operation. Observe the relations shown in Fig. 4-11. A semiconductor diode is forward-biased to 0.3 volt, and a small signal is applied from an audio generator. Since the AC signal has only a limited excursion on the diode characteristics, it is

88 · Basic Audio Amplifier Design

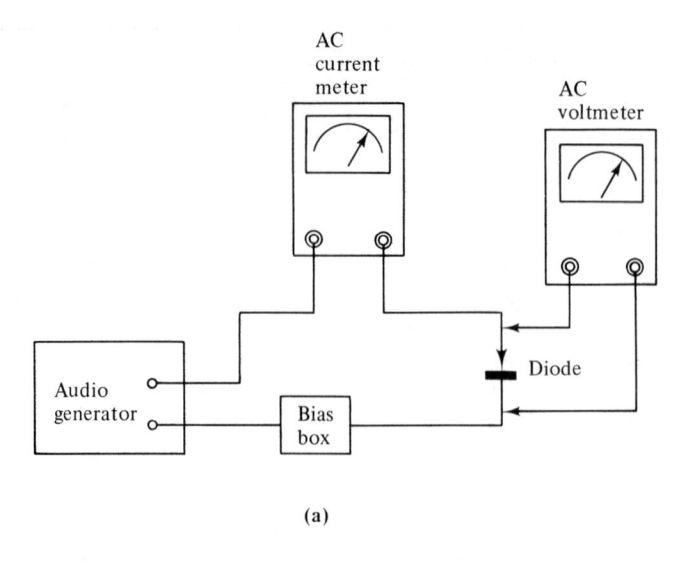

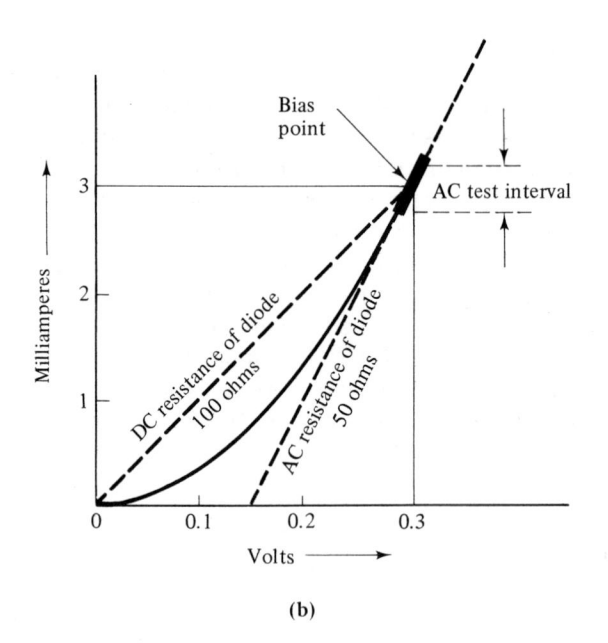

Figure 4–11 Comparison of AC resistance and DC resistance in a nonlinear system: **(a)** measurement of AC resistance; **(b)** AC resistance and DC resistance of a semiconductor diode.

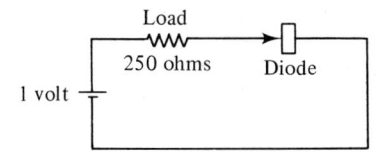

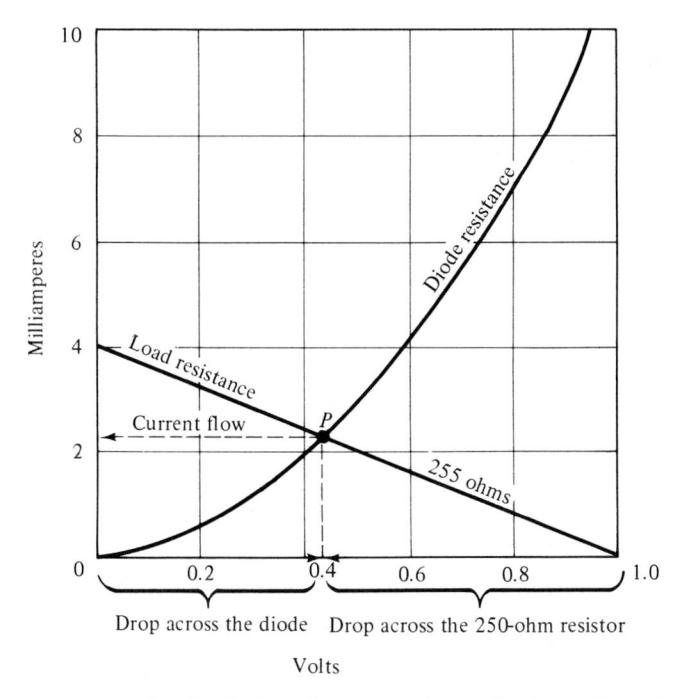

Figure 4–12 Graphical solution of current value and voltage drops in a diode circuit.

practical to assume that the diode has a linear resistance over this excursion. Note that the AC resistance of the diode is 50 ohms, whereas the DC resistance of the diode is 100 ohms in this example. In turn, if the AC current flow were calculated on the basis of the diode's DC resistance, *the answer would be 100 percent in error.* Therefore, the current flow must be calculated on the basis of the diode's AC resistance at its operating point (0.3 volt in this example).

Next, consider a simple circuit comprising both linear and nonlinear resistance, as exemplified in Fig. 4–12. Since the diode has a

90 · Basic Audio Amplifier Design

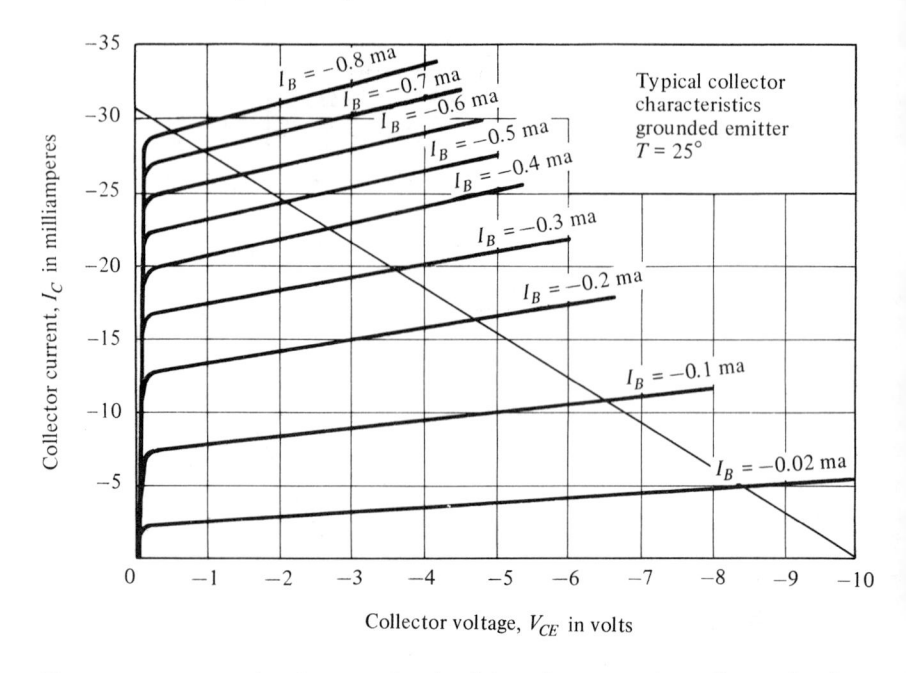

Figure 4-13 Example of a 333-ohm load line drawn on the collector family characteristics of a bipolar silicon transistor.

nonlinear resistance characteristic, it would be prohibitively difficult to calculate the operating point P. However, if a graph of the diode characteristic is utilized, the operating point is readily determined by drawing a *load line* on the plot, as shown in the diagram. The load line defines the resistance value of the linear resistor, in terms of its voltage drop and current flow. Thus, if the resistor drops 1 volt, the current flow through the circuit must be 4 mA; if the resistor drops zero volts, the current flow must be zero. These limiting values are indicated by the ends of the load line. The operating point occurs at the intersection of the load line with the diode characteristic. In this example, this point corresponds to a drop of 0.44 volt, approximately, across the diode, and a drop of 0.56 volt, approximately, across the linear resistor.

The same basic principles apply to a load line drawn on the collector family characteristics of a bipolar transistor, as shown in Fig. 4-13. In this example, $V_{cc} = 10$ volts, and $R_L = 333$ ohms. It is evident that the transistor is a nonlinear device, because the characteristics are expanded at low current levels, and are compressed at

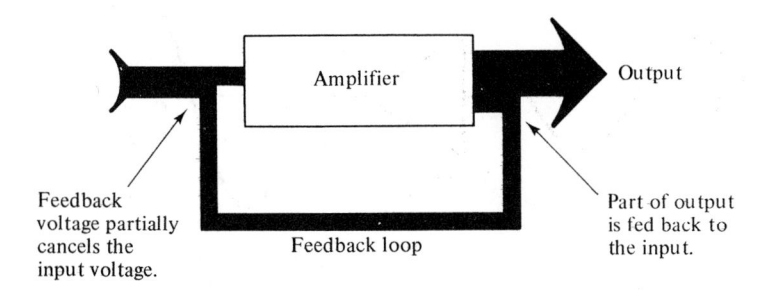

Figure 4–14 Basic principle of negative feedback.

high current levels. Suppose that the circuit designer chooses a base-bias current of 0.2 mA. Then, if the input signal swings the base ±0.1 mA, the current amplification of the stage will be 40 times. On the other hand, if the circuit designer chooses a base-bias current of 0.5 mA, and if the input signal swings the base ±0.1 mA, the current amplification of the stage will be 17.5 times, approximately. Thus, *the stage amplification will more than double when the lower-current operating point is used.*

4–8 ELEMENTS OF NEGATIVE FEEDBACK

To minimize the effects of inherent nonlinearity in a transistor, the circuit designer utilizes negative-feedback action. The basic principle of negative feedback is shown in Fig. 4–14. In other words, a portion of the output signal is fed back and mixed with the input signal so that partial cancellation of the source voltage occurs. Reduction of nonlinear distortion by negative feedback takes place, as seen in Fig. 4–15. Note that the mixture of the source waveform with the negative-feedback waveform serves to predistort the input signal to the transistor. This predistortion tends to compensate for the transistor distortion, so that the amplifier output signal has reduced distortion. If a sufficiently large amount of negative feedback is used, an amplifier can be made as nearly distortionless as desired. Note in passing that negative feedback does not make a transistor operate linearly—negative feedback merely compensates for the transistor distortion, so that the overall amplifier operation is effectively linearized.

Consider the circuit action in the negative-feedback amplifier arrangement shown in Fig. 4–16. Since a 10-mV input to the ampli-

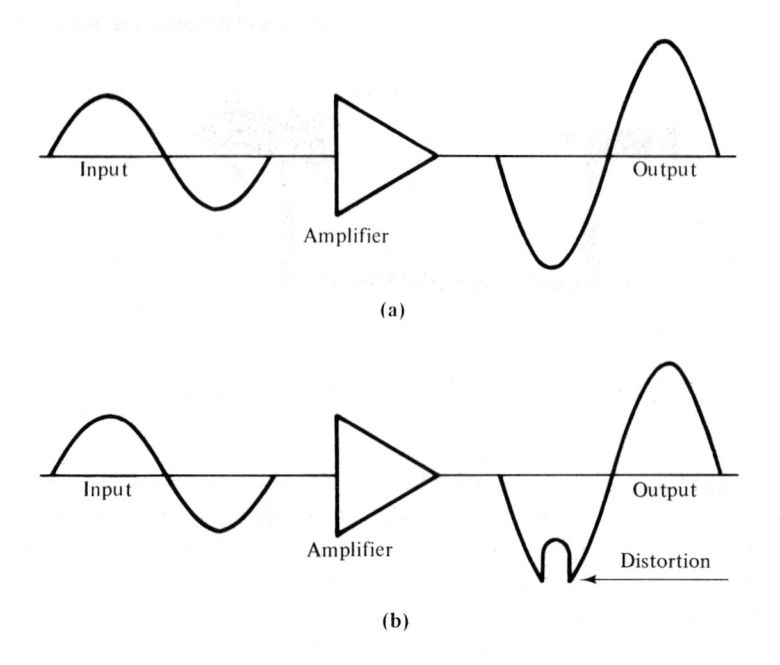

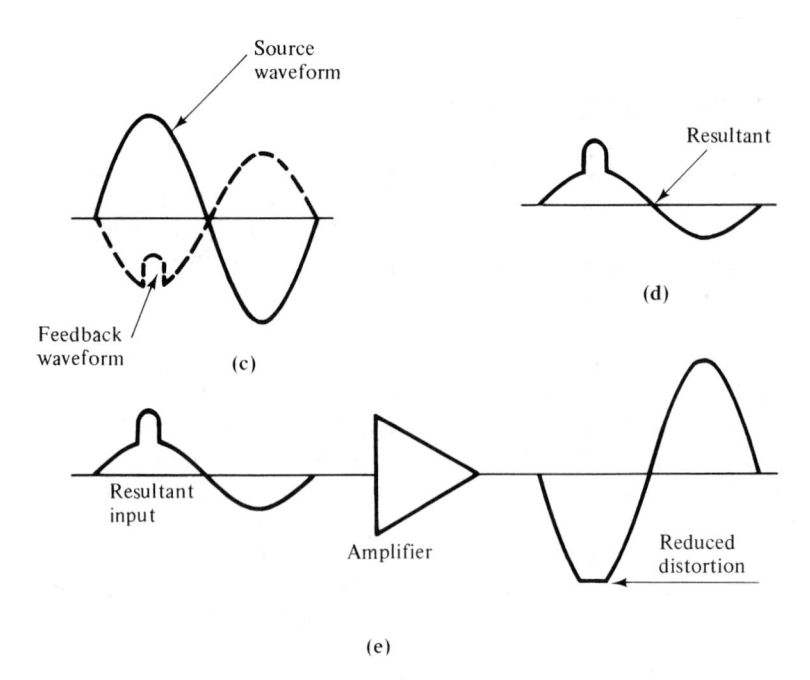

Figure 4-15 Amplifier operation with and without negative feedback: **(a)** ideal amplifier operation, with distortionless output; **(b)** amplifier output seriously distorted; **(c)** source waveform mixed with negative-feedback waveform; **(d)** resultant input waveform to amplifier; **(e)** amplifier output has reduced distortion.

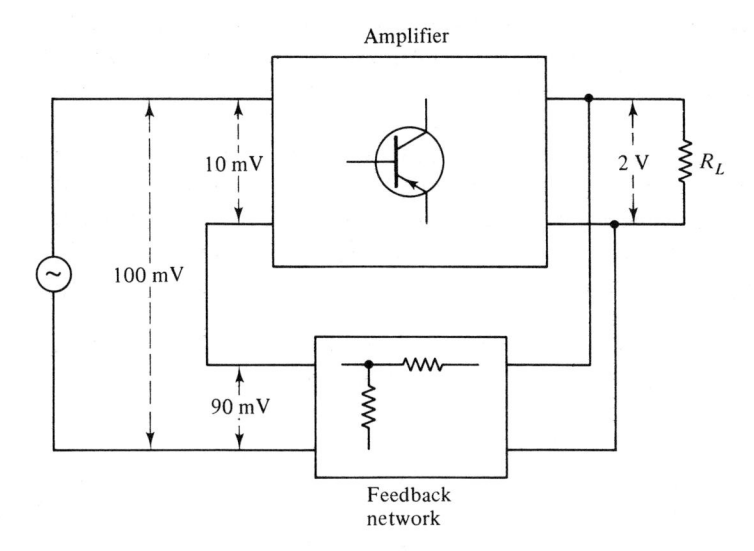

Figure 4–16 Example of negative-feedback circuit relations.

fier provides a 2-V output, the gain of the amplifier by itself is 200 times. Observe next that the output voltage drives the negative-feedback circuit. The output from the feedback network is 90 mV. This feedback voltage opposes the source voltage of 100 mV, leaving 10 mV at the input terminals of the amplifier. The end result is that a generator voltage of 100 mV produces 2 V output from the amplifier. In other words, the gain of the stage is 20 times. We can also say that the use of negative feedback in the stage reduces the gain from 200 to 20 times. This is a large reduction in gain. However, in a high-fidelity design, it represents an essential *trade-off* to minimize distortion, as was illustrated in Fig. 4–15.

Another benefit of negative feedback is that tolerances on transistors have less influence on amplifier action than if negative feedback were not used. For example, suppose that the transistor in Fig. 4–16 is replaced by another transistor that provides only one-half of the former device gain. If negative feedback were not used, the output voltage would drop to one-half of its former value. However, when the negative-feedback arrangement in Fig. 4–16 is used, the output voltage drops to 90 percent of its former value, as a bit of algebraic calculation will reveal. In other words, this *negative-feedback action has reduced a 50 percent tolerance on transistor beta to an effective tolerance of 10 percent.* This is a desirable

94 · Basic Audio Amplifier Design

trade-off in many circuit-design situations. For example, it helps to ensure that a production run of amplifiers will have a specified minimum gain value, regardless of wide tolerances on transistor beta values.

Another advantage of negative feedback is reduction of noise that is produced within the amplifier itself. For example, suppose that the transistor depicted in Fig. 4-16 produces noise pulses as a result of marginal collector-junction leakage. A noise pulse then appears at the output; at the same time, a cancellation pulse is fed back to the input of the transistor, and the disturbance in the output is greatly reduced. The noise reduction takes place in the general manner depicted in Fig. 4-15. Note in passing that the negative-feedback arrangement in Fig. 4-16 is said to have *significant feedback*. This term means that the amount of negative feedback that is employed serves to reduce the gain of the amplifier to 25 percent or less of its original value. Since the gain reduction in this example is to 10 percent of its original value, significant feedback action occurs. A voltage gain reduction to 10 percent is also called 20 dB of negative feedback.

Negative feedback also improves the frequency response of an amplifier. For example, suppose that the transistor in Fig. 4-6 has a beta cutoff frequency of 10 kHz. In turn, if no negative feedback is used, the frequency response of the amplifier is shown by the solid curve in Fig. 4-17. On the other hand, if 20 dB of negative feedback is utilized, the frequency response of the amplifier is shown by the dashed curve in Fig. 4-17. Thus, *a trade-off of 20 dB in gain provides a 2.7-dB improvement in gain of the transistor at its 3-dB cutoff point.* If the designer uses less negative feedback, the improvement in frequency response of the amplifier will be less; but if he employs more negative feedback, the improvement in amplifier frequency response will be greater than in the example cited. In any case, the improvement in frequency response can be calculated algebraically.

There are two basic negative-feedback arrangements, termed *voltage feedback* and *current feedback,* as shown in Fig. 4-18. Voltage feedback is provided by a resistor connected from collector to base of the transistor. Current feedback is provided by a resistor connected in series with the emitter lead. Note that voltage feedback decreases the output resistance of the amplifier. On the other hand, current feedback increases the output resistance of the amplifier. It is evident that both voltage feedback and current feedback operate by opposing the input voltage to the amplifier by a certain fraction of

Elements of Negative Feedback · 95

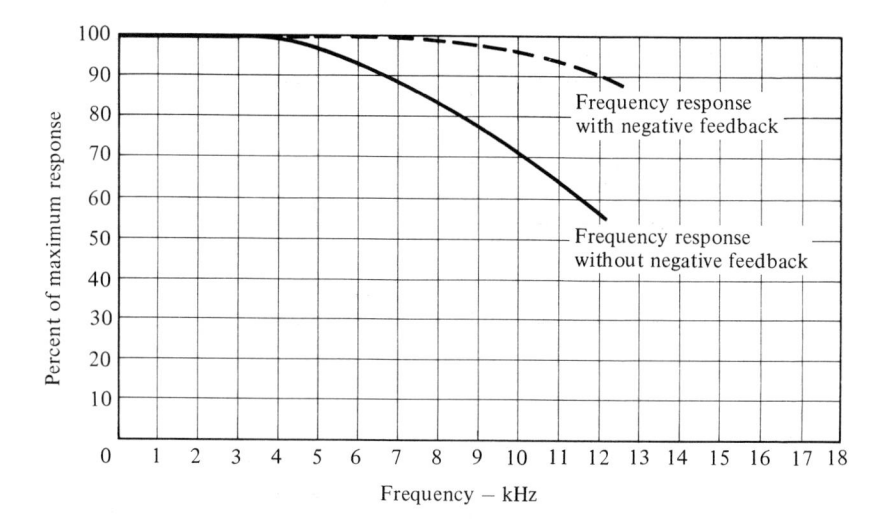

Figure 4-17 Improvement of frequency response provided by negative feedback.

the output voltage. The fraction of the output voltage that is mixed (in phase opposition) with the input voltage is often termed B. The amplification of the stage without negative feedback is often termed A. These terms are useful to calculate the approximate reduction in harmonic distortion that results from use of negative feedback, according to the equation

$$D_n = \frac{D_0}{1 + AB}$$

where D_n is the percentage distortion with negative feedback.
D_0 is the percentage distortion without negative feedback.
A and B are defined as noted above.

As an example of reduction in harmonic distortion, suppose that the amplifier depicted in Fig. 4-16 develops 5 percent of harmonic distortion without negative feedback. In this example, $A = 200$, and $B = 0.045$. In turn, the percentage of distortion with negative feedback will be approximately 0.5 percent. Thus, *20 dB of negative feedback provides a 90 percent reduction in harmonic distortion in this situation.* Observe in Fig. 4-16 that a generator voltage of 100 mV results in an output voltage of 2 volts from the stage. If no negative feedback is

96 · Basic Audio Amplifier Design

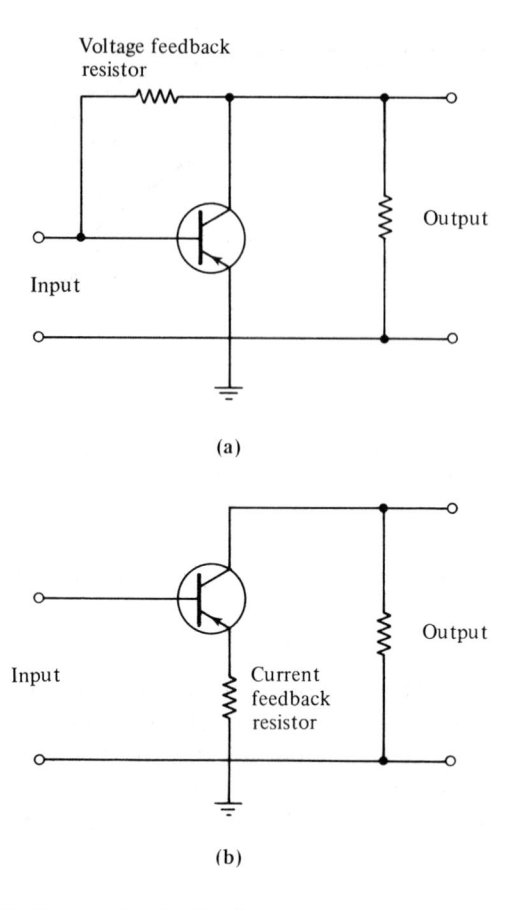

Figure 4-18 Basic negative-feedback arrangements: (a) voltage feedback; (b) current feedback.

used, a generator voltage of 10 mV will result in an output of 2 volts. Thus, when negative feedback is used in a stage, the input signal level may be increased to obtain the same output level as would prevail with no negative feedback. On the other hand, if the input signal level cannot be increased, or cannot be increased sufficiently to obtain the required output level, the circuit designer will cascade stages.

In addition to the advantages of negative feedback that have been noted, the tolerance on the value of the load resistance is also relaxed. With reference to Fig. 4-16, suppose that no negative feedback is used, and that the value of R_L is increased 20 percent. From a practical viewpoint, the output voltage will increase 20 percent as

Elements of Negative Feedback · 97

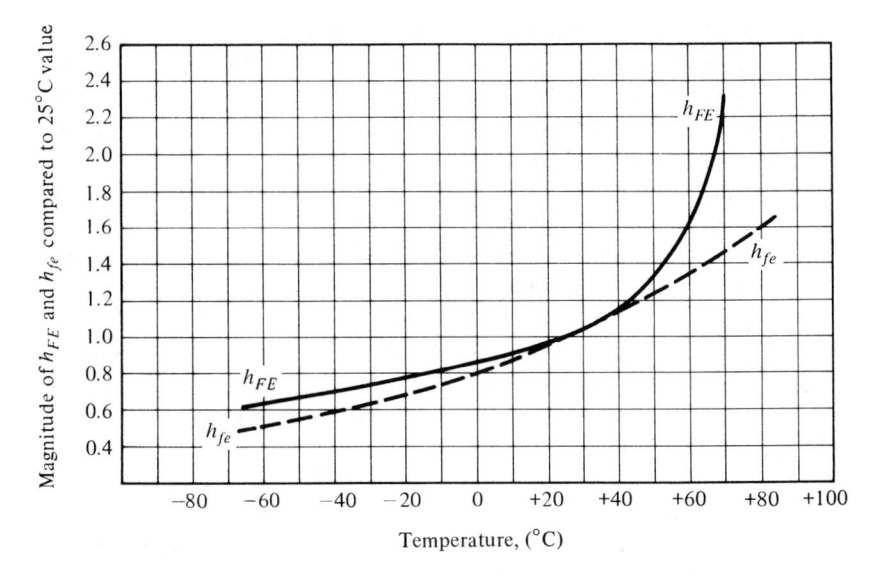

Figure 4-19 Variation in AC and DC beta values for a typical germanium transistor.

a result. That is, the output level will increase from 2 volts to 2.4 volts. On the other hand, when 20 dB of negative feedback is used, and the value of R_L is increased 20 percent, the output level will increase by only 1.5 percent, to a level of 2.03 volts, approximately.

Still another benefit of negative feedback is the effective stabilization of beta value that it provides for a transistor under conditions of temperature variation. As shown in Fig. 4-11, a semiconductor junction has an AC resistance value and a DC resistance value. Accordingly, a transistor has an AC beta value and a DC beta value. The former value is generally termed h_{fe}, and the latter h_{FE}. Variations in AC and DC beta values for a typical germanium transistor are shown in Fig. 4-19. In this topic, we consider the stabilization of h_{fe}; the stabilization of h_{FE} will be explained subsequently under bias-circuit design. Observe that the value of h_{fe} increases in the temperature range from $-55°$ to $+85°C$, and varies approximately 3 to 1. This variation is equivalent to a ± 50 percent tolerance on the value of beta. With reference to Fig. 4-16, it was shown that 20 dB of negative feedback will reduce this tolerance on beta to an effective tolerance of ± 10 percent.

It should be noted that *clipping distortion cannot be corrected by means of negative feedback*. The reason for this inability is de-

98 · Basic Audio Amplifier Design

picted in Fig. 4–20. Feedback of the clipped output waveform results in a predistortion of the input waveform to the amplifier. This predistortion consists of an increased peak voltage for the input waveform over the interval corresponding to clipping action in the amplifier. However, this increased peak voltage cannot affect the output from the amplifier, because there is no reserve gain beyond the clipping level. Accordingly, the predistorted input becomes clipped in the amplifier, and the distortion remains in the output waveform. A distinction should be made between *compression distortion* and *clipping distortion*. When a waveform is compressed, the amplifier still has some reserve gain, although its value is subnormal. On the other hand, when a waveform is clipped, the amplifier has zero reserve gain. Hence, compression can be reduced by negative-feedback action, whereas clipping cannot be corrected.

In addition to the extension of high-frequency response that results from the use of conventional negative feedback, additional extension can be provided by the use of a frequency-compensated negative-feedback loop, as shown in Fig. 4–21(a). Capacitor C has less reactance as the frequency increases. In turn, the amount of negative feedback decreases at higher frequencies and the amplifier gain increases. There is a trade-off involved in this arrangement, because the percentage of harmonic distortion will increase at higher frequencies. A limit on the amount of extension in the high-frequency response is imposed by the cutoff frequency characteristic of the transistor that is used. In other words, the transistor beta decreases to unity at some upper frequency; in turn, the amplifier high-frequency response is limited by this characteristic.

An alternative form of frequency-selective circuitry is seen in Fig. 4–21(b). This is not a feedback loop; it is an amplifier coupling circuit. Its frequency characteristics depend upon the input capacitance and resistance of the second stage. When R and C have suitable values, an extension of the amplifier's high-frequency response is obtained, owing to the decreasing reactance of C with increasing frequency. A trade-off is involved in this arrangement, in that the RC coupling network reduces the amplifier gain at lower frequencies. This coupling circuit and the foregoing frequency-compensated negative-feedback loop are basically RC filter sections, such as those discussed in Chapter 2.

Next, observe the frequency-independent RC voltage divider depicted in Fig. 4–21(c). This configuration is used in an amplifier network to reduce the signal level without introducing any frequency distortion. In other words, if the time constant of the first section is equal to the time constant of the second section, the output voltage will be a specified fraction of the input voltage at any frequency. Or

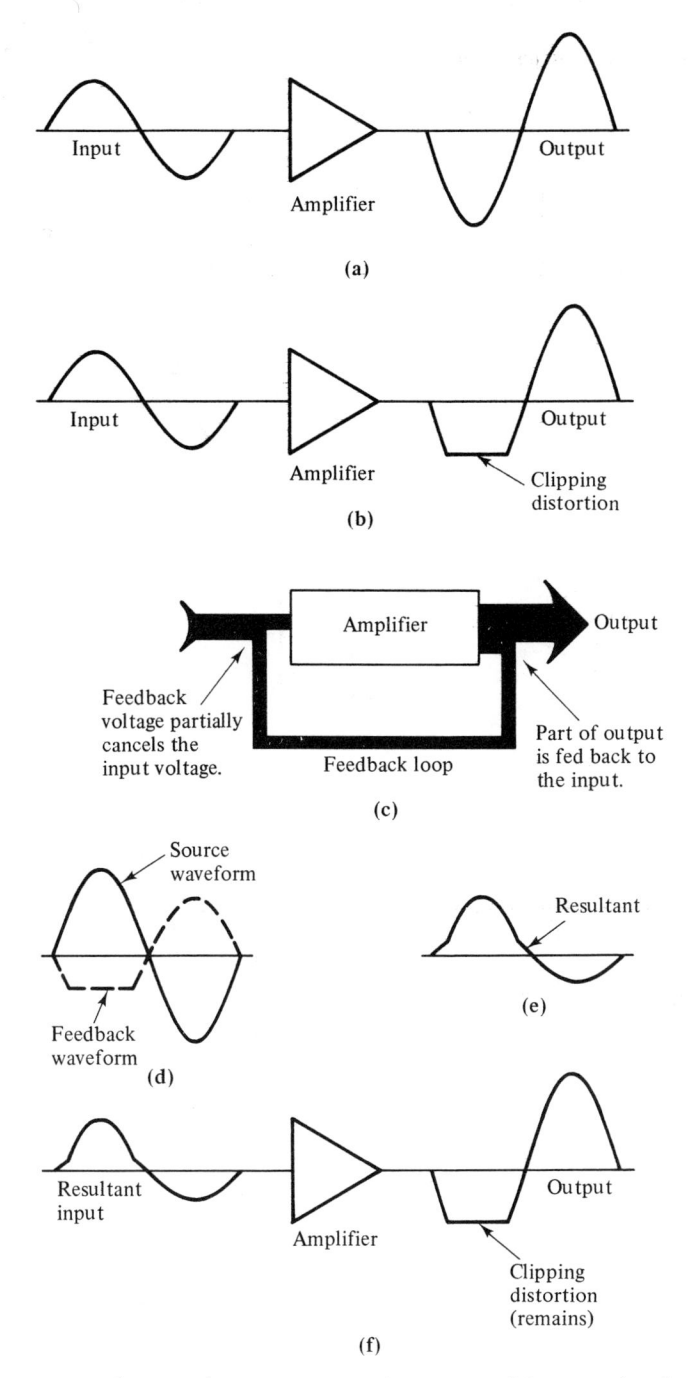

Figure 4–20 Clipping distortion cannot be corrected by negative feedback: **(a)** distortionless output; **(b)** clipped output; **(c)** negative-feedback arrangement; **(d)** source waveform mixed with feedback waveform; **(e)** resultant waveform; **(f)** resultant input waveform produces clipped output waveform.

99

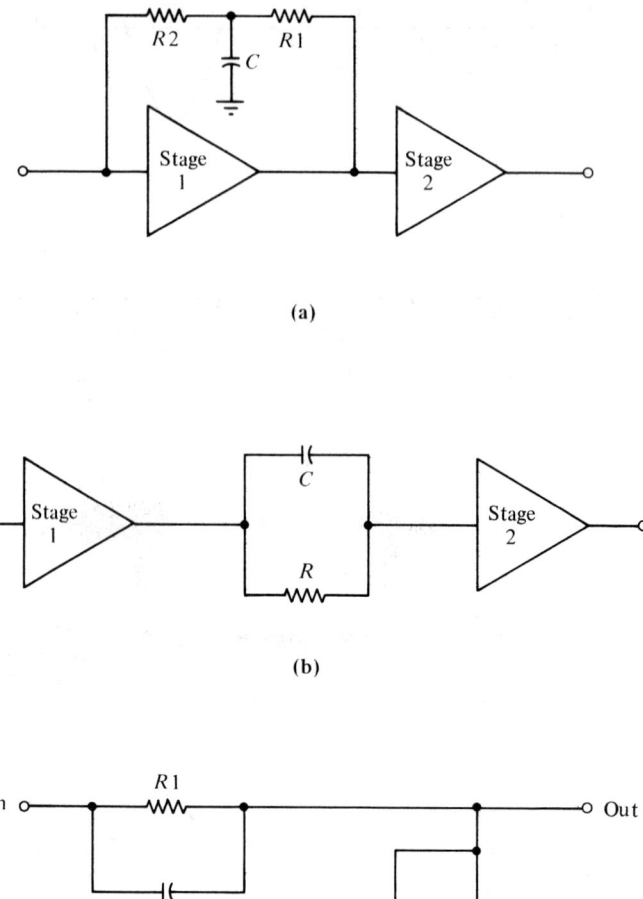

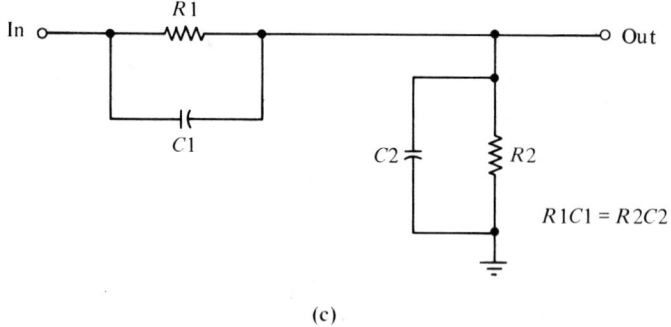

Figure 4–21 Frequency-selective and frequency-independent RC circuits: **(a)** frequency-selective negative-feedback loop; **(b)** frequency-selective amplifier coupling circuit; **(c)** frequency-independent RC voltage divider.

if a complex waveform such as a square wave is being processed by the amplifier, no waveform distortion will be imposed by the frequency-independent *RC* voltage divider. Considered as an *RC* filter arrangement, this configuration is called an *all-pass filter*.

4–9 ACTIVE TONE CONTROL

RC tone-control circuitry was discussed previously. Circuit designers often employ active tone-control arrangements, in which *RC* filter sections with adjustable resistance elements are included in the input and/or output branches of one or more amplifier stages. A two-stage arrangement with separate bass and treble tone controls is shown in Fig. 4–22, with functional diagrams for the bass-control and treble-control sections. Bass boost results from the bypassing of the higher audio frequencies by C2, and bass cut results from the attenuation of the lower audio frequencies by C1. Treble boost results from the action of C4, whereas treble cut results from the action of C5. The frequencies at which boost and cut responses occur depend upon the relative values of the capacitors.

In customary tone-control design, the capacitor values are chosen so that the treble control has little or no effect on bass response, and vice versa. Typical frequency-response curves for extreme settings of the tone controls are shown in Fig. 4–23. Although the configuration exemplified in Fig. 4–22 includes two transistors, its overall gain is practically unity, owing to the insertion losses of the *RC* filter sections. In other words, the active portion of the arrangement merely cancels out the insertion loss of the passive sections. *Tolerances* on *R* and *C* values can be analyzed on the basis of the functional diagrams in Fig. 4–22. That is, the *worst-case conditions* for *RC* filter circuitry explained in Chapter 2 can either be applied directly or applied in principle to these tone-control configurations.

An active tone-control configuration with junction field-effect transistors (JFET's) is shown in Fig. 4–24. Unlike the arrangement depicted in Fig. 4–22, this configuration does not use device isolation between the bass and treble *RC* sections. Interaction between subsections is extensive. Instead of analyzing the effect of tolerances on the network mathematically, it is more expedient in this kind of situation to make an experimental analysis. In other words, the circuit designer is advised to prepare a breadboard assembly at the outset. In turn, components with high-limit and low-limit tolerances can be plugged into the network, and their effects on circuit response

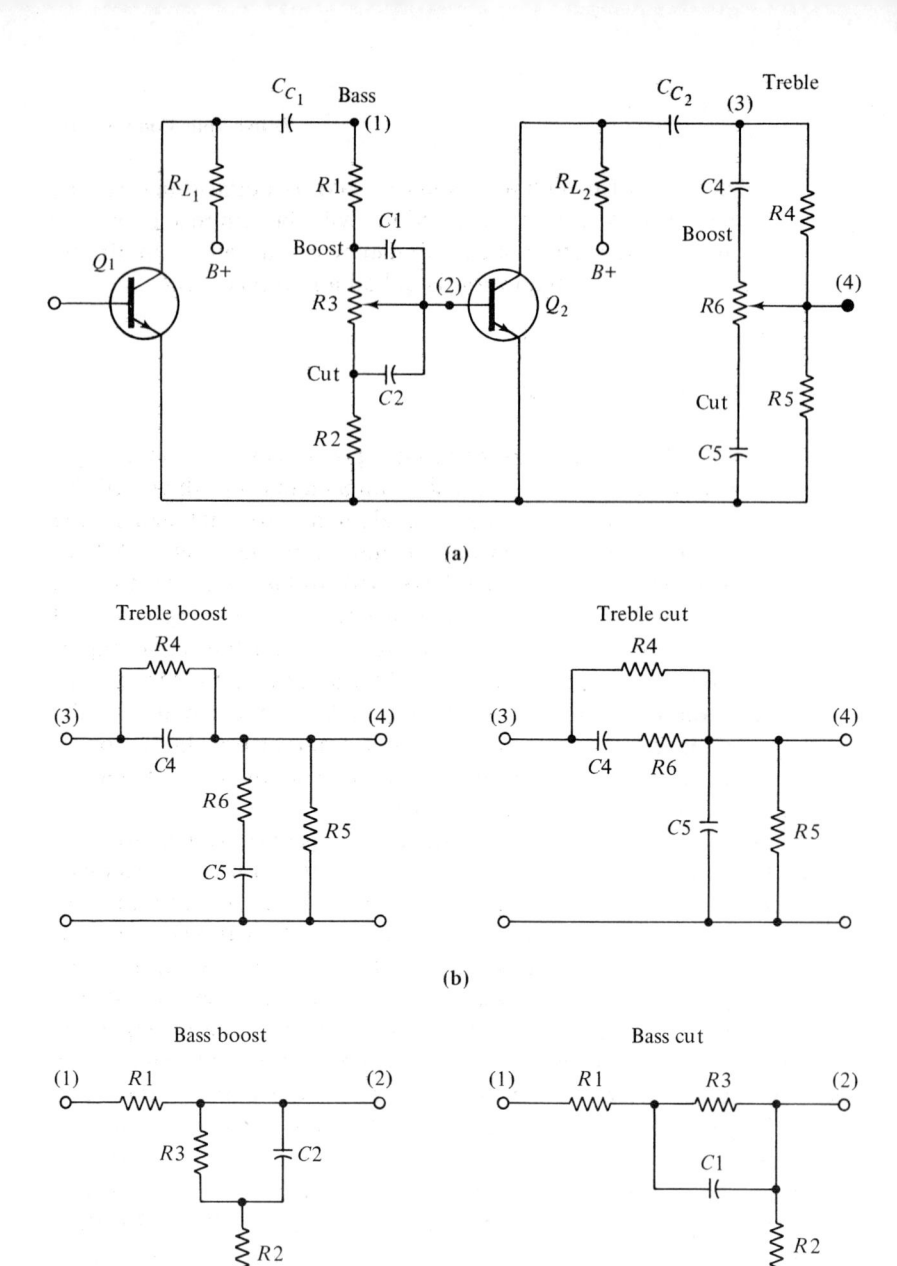

Figure 4–22 Typical configuration for an active bass-treble tone control: **(a)** bass/treble RC tone-control sections with device isolation; **(b)** treble section equivalent circuits at extreme control settings; **(c)** bass section equivalent circuits at extreme control settings.

102

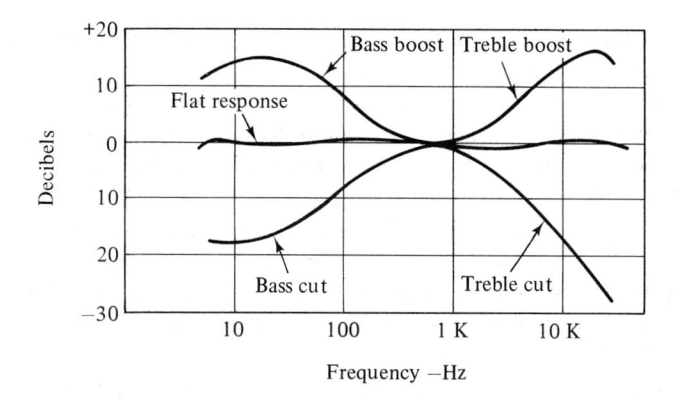

Figure 4-23 Frequency-response curves for extreme settings of a typical bass-treble tone-control arrangement.

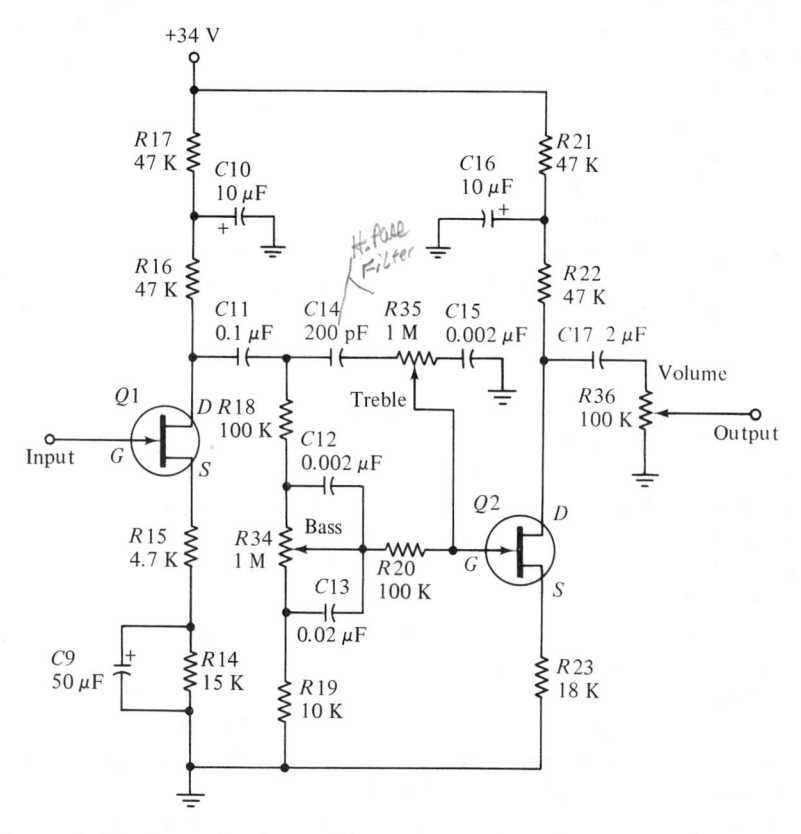

Figure 4-24 An active bass-treble tone-control configuration with junction field-effect transistors.

103

noted. Although considerable time is required to determine the worst-case conditions experimentally, still more time would ordinarily be required to make a corresponding mathematical analysis. Of course, if a large computer is available to the designer, the computer can be programmed to make the mathematical analysis.

Note that R23 is a source-bias resistor for Q2. It develops a voltage drop that determines the operating point for Q2. Because the resistor is not bypassed, it provides negative feedback for this stage, and improves the fidelity (linearity) of response. Resistor R22 is the drain load, across which the output signal is developed. Decoupling is provided by C16/R21. Self-bias for Q1 is provided by the source-bias resistors R14 and R15. Only R15 provides negative feedback, because R14 is bypassed. In turn, less gain is traded off for improved fidelity of response in the Q1 stage than in the Q2 stage. Observe that in the absence of negative feedback, each stage in Fig. 4–24 will develop a typical gain of 40 times. However, the effect of degeneration owing to an unbypassed source resistor reduces the stage gain according to the equation

$$\frac{A_{fb}}{A} \approx \frac{1}{1 + AB}$$

where A_{fb} is the stage amplification with degeneration.
A is the stage amplification without degeneration.
B is the ratio of the source resistance to the sum of the source and drain resistances.

Accordingly, if the Q1 stage in Fig. 4–24 has a gain of 40 times when the source resistors are bypassed, it will provide a gain of approximately 8.8 times when R15 is unbypassed. Similarly, if the Q2 stage in Fig. 4–24 has a gain of 40 times when the source resistor is bypassed, it will provide a gain of approximately 3.3 times when R23 is unbypassed. To calculate the stage gain with the source resistor bypassed, multiply the transconductance of the JFET by the value of the drain (load) resistance. Note that in the case of the Q2 stage, the 47,000-ohm drain resistor is effectively connected in parallel with the 100,000-ohm volume control. Hence, the designer will calculate the stage gain on the basis of a 32,000-ohm drain-load resistor. Bogie transconductance values for JFET's range up to 2000 μmhos, or more, and are published in manufacturers' data sheets. These same design principles apply to MOSFET amplifiers.

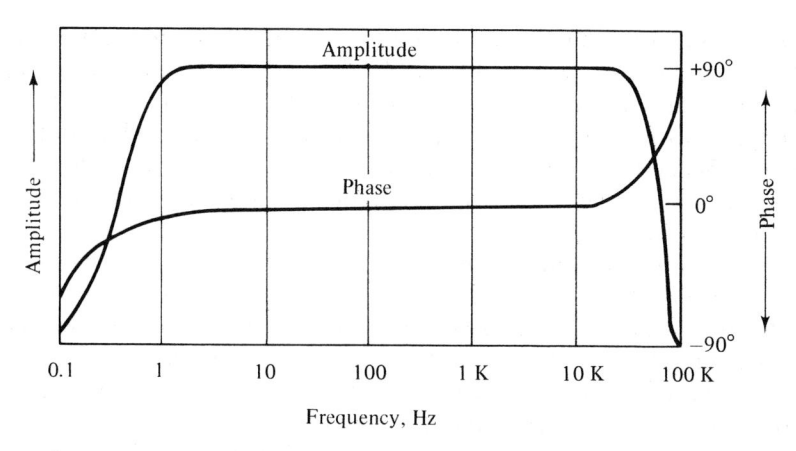

Figure 4–25 Amplitude and phase characteristics for a hi-fi amplifier.

4–10 MULTISTAGE NEGATIVE-FEEDBACK HAZARD

Single-stage negative feedback is in very wide use. Degeneration resulting from an unbypassed emitter resistor is most common; voltage feedback from collector to base is also utilized in many designs. Two-stage negative feedback is less usual, although it is in fairly extensive use. For example, voltage feedback can be employed from the collector of the second-stage transistor to the emitter of the first-stage transistor. On the other hand, three-stage negative feedback is rarely utilized, because of the inherent instability hazard. Although it may appear superficially feasible to use voltage feedback from the collector of the third-stage transistor to the base of the first-stage transistor, amplifier phase shift is likely to introduce instability and the possibility of oscillation in the form of motorboating or squegging.

A single stage of amplification has a phase characteristic similar to that shown in Fig. 4–25. Zero phase shift occurs only at the mid-band frequency. As the frequency characteristic passes through cutoff, either at the high end or at the low end of its range, the phase characteristic varies rapidly and approaches 90 deg lagging or leading as a limit. When two stages are connected in cascade, the phase characteristic approaches 180 deg as a limit at the extreme ends of its frequency range. Or when three stages are connected in cascade, the phase characteristic approaches 270 deg as a limit. Note that when the phase shift is 180 deg, the *feedback is positive,* instead of negative. In other words, the 180-deg phase shift provided by a common-emitter stage is cancelled by the feedback voltage that has been shifted in phase by 180 deg. In turn, positive feedback permits the amplifier to oscillate. In summary, whenever an amplifier can supply its own input at a level that sustains the prevailing output, the amplifier breaks into oscillation. This instability may occur either at the low-frequency end of its frequency range, or at its high-frequency end.

Although a prototype model of an amplifier that utilizes multistage feedback operates in a stable manner, there is no assurance that the design will remain stable when transistors with higher beta values or with higher beta cutoff values are used in production. Similarly, the amplifier may become unstable when resistors and capacitors with perhaps a ±20 percent spread of values are used in production. Accordingly, whenever negative feedback is used over more than one stage, *the circuit designer should make an intensive worst-case analysis of potential instability owing to cumulative phase shifts.* This worst-case analysis must take temperature variations and supply-voltage variations into consideration. If there is a possibility that the amplifier may be used with an inductive or a capacitive load, instead of a purely resistive load, this potential source of instability should be taken into account. Similarly, if the amplifier may be used with a generator that has a substantially different value of internal resistance than is used to test the prototype model, this operating parameter should be checked.

Transistor Worst-case Factors To make a worst-case analysis of any amplifier circuit design, the following transistor characteristics, limitations, and operating factors should be taken into consideration:

1. *Deterioration with time.* Conservative ratings of transistor parameters should be assigned to allow for reasonable deterioration in operating characteristics with time. This allowance is of particular importance in high-reliability designs and in uncompromising design procedures.
2. *Current gain versus temperature.* At collector-current levels above a few mA, the alpha or beta value of a transistor decreases at a greater rate as the temperature increases.
3. *Power dissipation versus temperature.* Circuit-design provisions may be required to avoid excessive power dissipation by the transistor at higher temperatures. When breadboard worst-case analyses are made, it is advisable to include a

fuse or a current-limiting resistor to avoid accidental damage to the transistor owing to excessive collector-current flow.
4. *Frequency limitations.* Frequency cutoff limits in common-emitter or common-collector configurations depend on forward current-gain values. Collector capacitance is a contributing factor in the high-frequency response limit.
5. *Collector leakage currents.* Collector leakage current will double for each 8-deg to 10-deg increase in operating temperature. In the common-emitter configuration, I_{CE} can vary from I_{CBO} to $h_{FE} \times I_{CBO}$. It is good practice to minimize the circuit resistance between base and emitter, insofar as stage gain is not affected adversely.
6. *Manufacturers' ratings.* Second-source considerations should be taken into account; different manufacturers may publish somewhat different ratings for the same transistor type number.
7. *Power considerations.* Apply a suitable thermal derating factor for operating temperatures above 25°C. This is approximately 1 to 10 mW/°C for small-signal transistors, and 0.25 to 1.5 W/°C for power-type transistors. Remember that the thermal resistance (R_T, Θ_R) in a manufacturer's rating does not include the thermal resistance of a heat sink. Include an emitter swamping resistance whenever this is practical, to avoid the possibility of thermal runaway.
8. *Temperature considerations.* Excessive leakage currents can be avoided by limiting the maximum junction temperature. Bias variation can be reduced by limiting the minimum junction temperature; both germanium and silicon transistors have a negative temperature coefficient of 2 mV/°C. Large values of collector current (I_c) assist in reduction of collector-current variation owing to temperature changes. Low values of source resistance in driving a base circuit contribute to a low stability factor. When diodes are used for bias stabilization of a transistor, use a common heat sink for the devices. Conservative design employs a minimum f_{FE} value over the operating temperature range. Note that *a low stability factor will not improve the performance of a DC amplifier.*
9. *Operating voltage precautions.* Do not exceed V_{CB} maximum, V_{CE} maximum, or V_{BE} maximum (reverse breakdown voltage) ratings under any condition of circuit operation. In

a push-pull arrangement, V_{CC} should be less than half of V_{CB}. Be on guard for possible transient voltages (such as switching transients) that could exceed maximum ratings. Keep in mind that the reverse breakdown voltage of a silicon transistor decreases with increasing temperature.

10. *Soldering precautions.* Transistor damage in production operations can result from excessive heat during soldering. In a dip-soldering process during assembly of printed circuits with transistors, the temperature of the solder should not exceed 250°C for a maximum immersion period of 10 seconds. The flange of a transistor must not be soldered to a heat sink.

Production Cost Trade-offs As a rough rule of thumb, a class-A audio amplifier with a rated power output of less than 0.1 watt will be least expensive in production if it is designed with direct-coupled input and direct-coupled output circuits. However, it is more difficult to realize good bias stability versus temperature changes than if transformer coupling were utilized. Again, an audio amplifier with a rated power output of more than 0.1 watt will be least expensive in production if it is designed with transformer-coupled output and direct-coupled input. On the other hand, the bias stability of this arrangement versus temperature changes is poorer than if input transformer coupling were employed. Note also that transformer coupling introduces more harmonic distortion than direct coupling. Another trade-off that the circuit designer must consider is the comparatively limited frequency range inherent in transformer coupling.

CHAPTER 5

Amplifier Bias Stabilization

5–1 GENERAL CONSIDERATIONS

Most of the voltages in an amplifier configuration have rather wide tolerances. However, *transistor bias voltages are subject to unusually tight tolerances*. The necessity for close control of the base-emitter bias voltage is seen in Fig. 5–1. A very small change in forward bias voltage can cause a very large change in forward current flow. In other words, the operating point of a transistor is critical. In turn, the circuit designer must use bias circuitry that maintains a tight tolerance on the bias voltage, regardless of temperature variation, supply voltage variation, and tolerances on various characteristics of replacement transistors.

Bias corresponding to the operating or quiescent point for a transistor is established by specifying quiescent values of collector voltage and emitter current. These are DC values of voltage and current, with no applied signal. Over a wide range of temperature, reliable operation of a transistor requires stability of bias voltage and current values. Compensating circuits are utilized to obtain bias stability under conditions of variation in reverse-bias collector current and variation in emitter-base junction resistance. Reverse-bias collector current is also called *saturation current*. Variation of saturation current with temperature of the base-collector junction is exemplified in Fig. 5–2. This variation in saturation current value extends from almost zero at 10°C to considerably over 1 mA at

110 · Amplifier Bias Stabilization

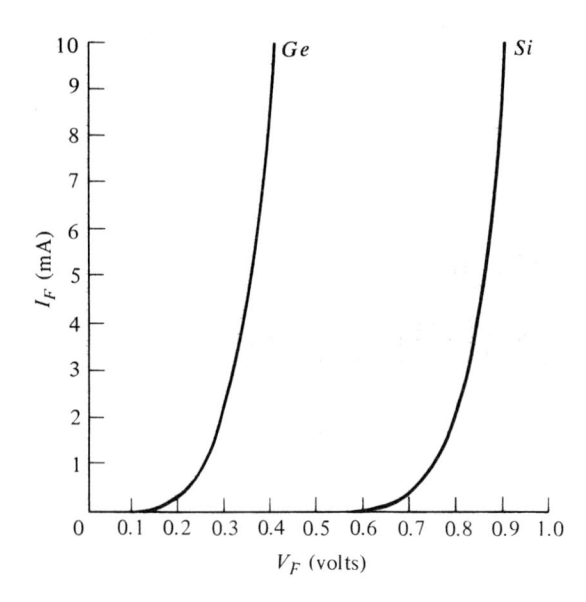

Figure 5-1 Forward current flow versus forward bias voltage for typical base-emitter junctions.

125°C. *At temperatures below 10°C, saturation current has no effect on bias.*

Note that Fig. 5-1 shows the variation in forward current flow versus base-emitter bias voltage, whereas Fig. 5-2 depicts the variation in saturation (reverse current) flow versus temperature. Saturation current increases very rapidly as the temperature increases and, as will be shown, causes a rapid increase in emitter current, with a concomitant rapid increase in collector current. Therefore, bias circuitry must be designed to stabilize the collector current under conditions of temperature variation and related variation in saturation current. This design feature has an additional advantage, in that *it makes the amplifier relatively independent of tolerances on saturation current of replacement transistors.*

Saturation current, in the case of an NPN transistor, consists of a flow of holes from the collector into the base. If resistors connected to the base have a high value, holes can accumulate in the base region. This accumulation causes an increase of emitter current (electrons) into the base and increases the collector current. In turn, the temperature of the collector-base junction is increased, and the saturation current value increases. This device action will continue until the amplifier distorts seriously; sometimes this "snowballing"

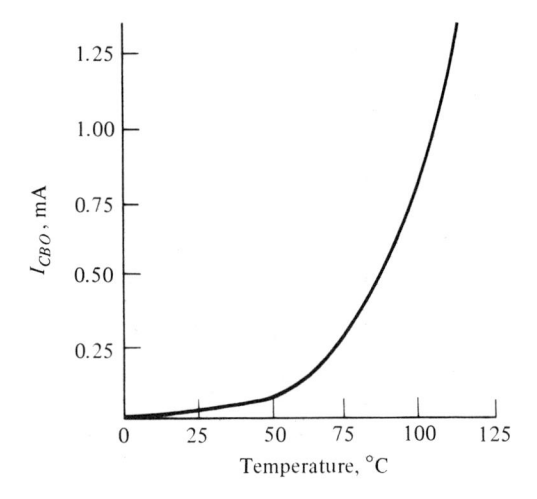

Figure 5-2 Variation of saturation current with junction temperature.

attains a level such that the transistor becomes inoperative, or possibly destroyed. *This hazard is minimized by avoiding the use of high-valued resistors in the base lead.* Similarly, in the case of a PNP transistor, "snowballing" occurs in the same way, except that roles of electrons and holes are interchanged.

Next, consider the relation of emitter-base junction resistance to the problem of bias stability. Variation of collector current with temperature is exemplified in Fig. 5-3. This is a typical *family* of curves. Each curve is plotted with a *fixed* collector-base voltage (V_{CB}) and a *fixed* emitter-base voltage (V_{EB}). It is evident that if collector current variation versus temperature were caused only by the action of saturation current, then the collector current variation at temperatures below 10°C ($V_{EB} = 200$ mV and $V_{EB} = 300$ mV) should not occur. However, the collector current *does* vary with temperature, even when the saturation current is near zero. This variation is caused by the decrease in emitter-base junction resistance as the temperature *increases*. In other words, *the emitter-base junction resistance has a negative temperature coefficient of resistance.*

One practical method of reducing the effect of this negative temperature coefficient of resistance is to connect a high value of resistance in series with the emitter lead. Effectively, this resistor causes the variation of the emitter-base junction resistance to be a small percentage of the *total* resistance in the emitter circuit. This external resistor *swamps* (overcomes) the junction resistance; hence, it is often

112 · Amplifier Bias Stabilization

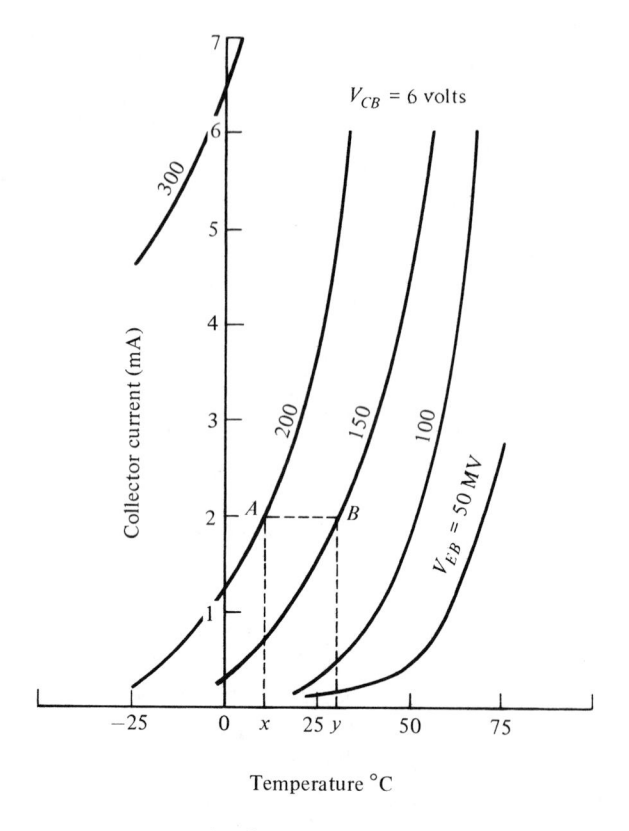

Figure 5-3 Variation of collector current with transistor temperature.

termed a *swamping resistor*. Another practical method of reducing the effect of this negative temperature coefficient of resistance is to reduce the emitter-base forward bias as the temperature increases. As an illustration, to maintain the collector current at 2 mA in Fig. 5-3 while the transistor temperature varies from 10°C (at X) to 30°C (at Y), the forward bias must be reduced from 200 mV (at A) to 150 mV (at B). The temperature difference is 20°C (30 − 10); the voltage difference (200 mV − 150 mV) is 50 mV. Accordingly, the variation in forward bias per degree Celsius is calculated as follows:

$$\frac{\text{Difference in forward bias}}{\text{Difference in temperature}} = \frac{50 \text{ mV}}{20°C} = 2.5 \text{ mV}/°C$$

The foregoing calculation indicates that the collector current will not vary with change in emitter-base junction resistance if the forward bias is reduced 2.5 mV/°C for increasing temperature, or if it is increased 2.5 mV/°C for decreasing temperature.

Implementation of Bias Stabilizing Circuits Bias stabilizing circuits may employ a variety of circuit elements, from which the circuit designer may make a judicious choice:

1. Common resistors, as explained in the next topic. The effectiveness of external resistors in bias stabilization of transistor amplifiers is indicated by an analysis of the general bias circuit, as will be demonstrated. This analysis results in mathematical expressions for current and voltage stability factors (S_I and S_V) for each amplifier configuration.
2. Thermistors, as discussed under a following topic.
3. Junction diodes, as explained subsequently.
4. Transistors, as detailed below.
5. Breakdown diodes, as described in a later topic.

5–2 RESISTOR STABILIZING CIRCUITS

It is helpful to consider a *general bias circuit*. Common-emitter, common-base, and common-collector configurations were noted previously. These are special cases of the general transistor circuit shown in Fig. 5–4(a). Depending upon the points of input and output of the signal into the general transistor circuit and the values assigned to the resistors, the three basic configurations (CE, CB, and CC) can be derived.

1. *CB Amplifier.* The CB amplifier depicted in Fig. 5–4(b) can be derived from the general transistor circuit by opening resistor R_F ($R_F = \infty$), and short-circuiting resistor R_B ($R_B = 0$). The input signal is introduced into the emitter-base circuit and is extracted from the collector-base circuit.
2. *CE Amplifier.* The CE amplifier, depicted in Fig. 5–4(c), can be derived by opening resistor R_F and short-circuiting resistor R_E. In this case, the signal is introduced into the base-emitter circuit, and is extracted from the collector-emitter circuit.
3. *CC Amplifier.* The CC amplifier, depicted in Fig. 5–4(d), can be derived by opening resistor R_F and short-circuiting

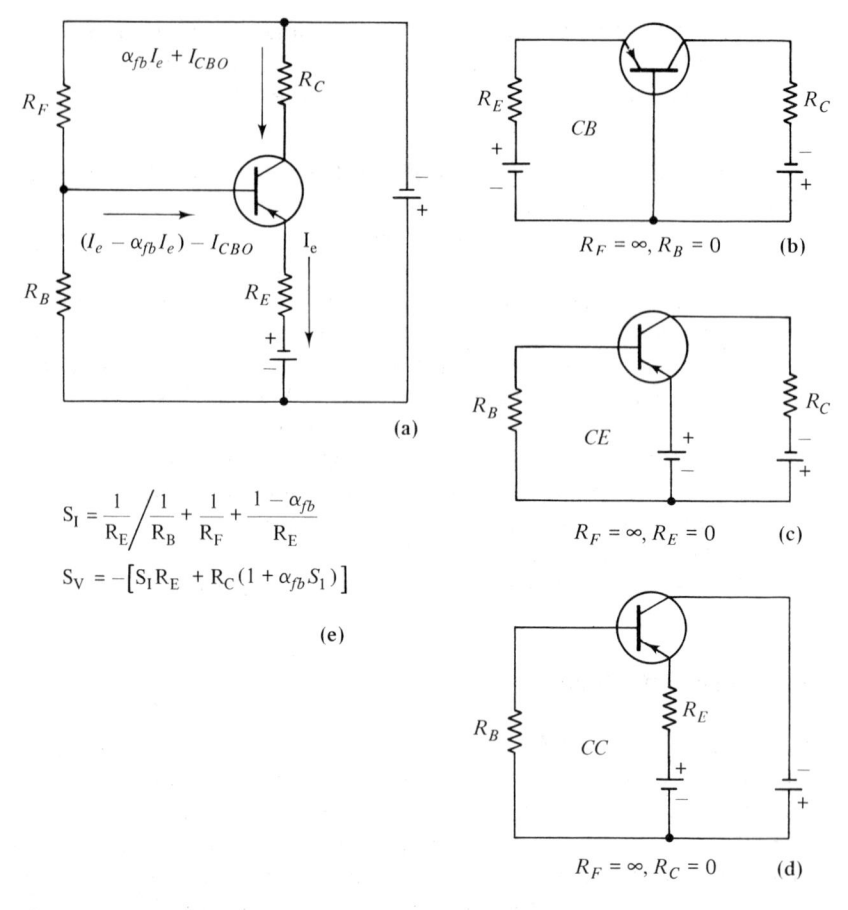

Figure 5–4 General transistor circuit and derived configurations, with current and voltage stability factors.

$$S_I = \frac{1}{R_E} \Big/ \frac{1}{R_B} + \frac{1}{R_F} + \frac{1-\alpha_{fb}}{R_E}$$

$$S_V = -\big[S_I R_E + R_C(1 + \alpha_{fb} S_1)\big]$$

(e)

resistor R_C. The collector is common to the input and output circuits. (Circuits in which R_F is not open will be described subsequently.)

Current Stability Factor Next, consider the current stability factor, with reference to Fig. 5–4(e).

1. The ratio of a *change in emitter current* ΔI_E *to a change in saturation current* ΔI_{CBO} is a measure of the bias stability of the transistor. This ratio indicates the effect on the emitter

current of a change in saturation current and is called the current stability factor S_I. It is expressed by the following equation:

$$S_I = \frac{\Delta I_E}{\Delta I_{\text{CBO}}}$$

Because the current stability factor is a ratio of two currents, it is expressed as a pure number. **Under ideal conditions, the current stability factor would be equal to zero;** in other words, the emitter current would not be affected by a change in the saturation current.

2. By using the resistors and the indicated expressions for current flow in Fig. 5-3(a), current I_E can be expressed in terms of saturation current I_{CBO}, and the resistors used in the circuit. Analysis of such an equation to indicate how current I_E varies with saturation current I_{CBO} results in the equation for the current stability factor S_I noted in Fig. 5-4(e). *This current stability factor defines the effectiveness of the external circuit to minimize emitter current variations.*

3. To derive the CB amplifier configuration from the general transistor circuit in Fig. 5-4, resistor R_F is opened ($=\infty$) and resistor R_B is short-circuited. To obtain the stability factor for this CB circuit, infinity is substituted for R_F and zero is substituted for R_B in the general equation for S_I. Note that the reciprocal of infinity is zero, and that the reciprocal of zero is infinity. In turn, S_I becomes zero; *this is the ideal condition,* inasmuch as it minimizes the accumulation of minority carriers in the base region of the transistor and permits the use of an emitter swamping resistor.

4. Directly substituting for the CE configuration depicted in Fig. 5-4(c) the values of resistor R_F ($=\infty$) and resistor R_E ($=0$) in the general equation for S_I gives an indeterminate answer of infinity divided by infinity. However, if we multiply the numerator and denominator by R_E before substitution, the value of the current stability factor is found:

$$S_I = \frac{1}{1 - \alpha_{fb}}$$

Because α_{fb} is approximately equal to unity, the value of S_I is high, meaning that *the inherent bias-current stability of the*

CE configuration is poor. This result could have been anticipated, because there is no swamping resistor present, and resistance is present in the base lead.

5. Substituting for the CC configuration depicted in Fig. 5–4(d) the values $R_F = \infty$, the approximate value of the current stability factor is expressed:

$$S_I = \frac{R_B}{R_E}$$

In other words, the value of S_I depends on the ratio of base resistance to emitter resistance. This result is a restatement of the basic principle that the higher the base resistance, the poorer is the current stability factor, and that the higher the emitter resistance, the better is the current stability factor.

Base Resistance and Emitter Resistance The effect on collector current stability of various combinations of base and emitter resistance values is exemplified in Fig. 5–5. Note carefully that the *worst-case condition* occurs when the emitter and base resistances are both zero. A slight improvement occurs if the base resistance is

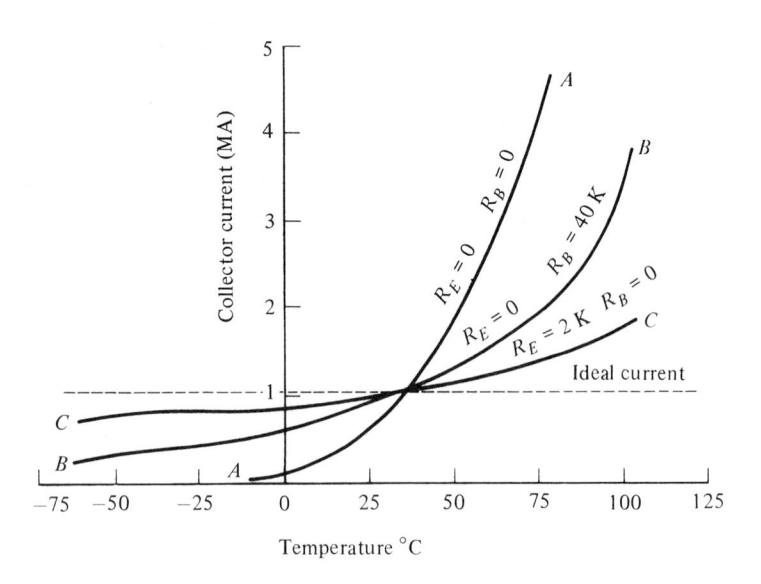

Figure 5–5 Variation of collector current with temperature, for different values of emitter resistance and base resistance.

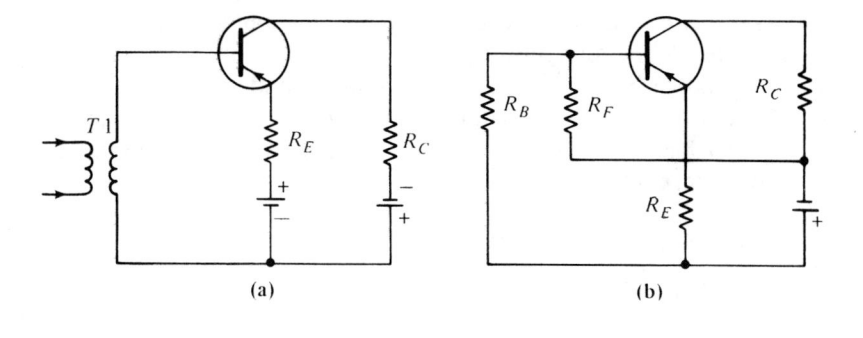

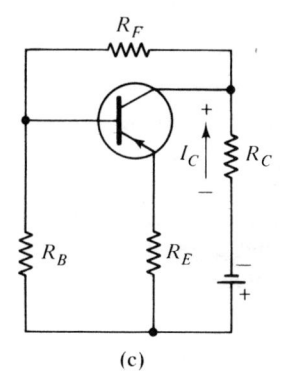

Figure 5–6 CE circuits that employ transformer input, fixed bias voltage, and negative feedback.

equal to 40,000 ohms and the emitter resistance is equal to zero. On the other hand, the *optimum condition* occurs when the emitter resistance is greater than zero (2000 ohms in this case), and if the base resistance is zero.

Additional Bias Techniques A practical method of obtaining good current stability in the CE configuration is to use near-zero base resistance and an emitter swamping resistor, as shown in Fig. 5–6(a). Resistor R_E operates as a swamping resistor; the secondary winding of the transformer offers a very low DC resistance in the base circuit. The collector current stability curve for this arrangement is similar to that of curve CC in Fig. 5–5. *This circuit has a current stability factor equal to zero.*

118 · Amplifier Bias Stabilization

Next, note that a fixed emitter-base bias can be obtained in a single-battery CE configuration by means of a voltage-divider network, as shown in Fig. 5–6(b). This voltage divider comprises resistors R_F and R_B. The voltage developed across R_B contributes part of the emitter-base voltage. Substitution in the general current-stability equation, Fig. 5–4(e), yields the expression

$$S_I = \frac{R_B R_F}{R_B + R_F} \bigg/ R_E$$

This equation states that the current stability factor for the arrangement in Fig. 5–6(b) is equal to the ratio of the parallel resistance value of R_B and R_F (base ground-return resistance) to the emitter resistance. Accordingly, this equation again substantiates the basic principle that the lower the base ground-return resistance, and the higher the emitter resistance, the better is the current stability.

Consider next a circuit that employs *negative feedback voltage* to improve the current stability, as exemplified in Fig. 5–6(c). Observe that if the collector current I_C rises, the collector becomes less negative because of the larger DC voltage drop across R_C. In turn, less forward bias (negative base to positive emitter) is coupled through resistor R_F to the base. Reduced forward bias results in reduced collector current.

5–3 VOLTAGE STABILITY FACTOR

The ratio of an *increment in collector voltage to an increment in saturation current* is a measure of the collector voltage stability of a transistor. This ratio indicates the effect on the collector voltage of a change in saturation current, and is called the *voltage stability factor*. It is expressed basically by the following equation:

$$S_V = \frac{\Delta V_{CB}}{\Delta I_{CBO}}$$

Because the voltage stability factor is a ratio of a voltage and a current, the corresponding unit is *resistance*. **Under ideal conditions, the voltage stability factor would be equal to zero.** In other words, the collector voltage would not be affected by a change in saturation current. The general voltage-stability factor is given in Fig. 5–4(e). Note that if the current-stability factor is zero, the voltage-stability

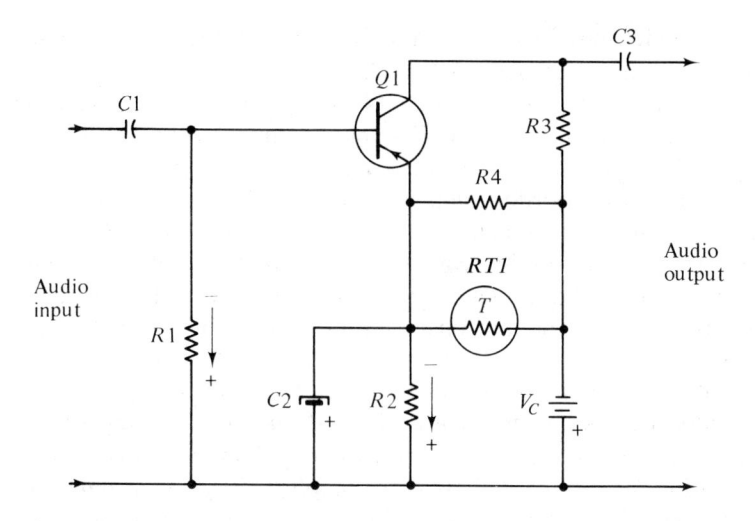

Figure 5–7 Typical transistor amplifier stage with thermistor control of emitter bias voltage.

factor equals the collector resistance. Use of a transformer with a low-resistance primary winding in the collector circuit tends to reduce the voltage stability factor to near zero. Practical circuits are discussed in a following topic.

5–4 THERMISTOR STABILIZING CIRCUITS

As developed in the foregoing topics, the bias current of a transistor is temperature-sensitive. In particular, the emitter current increases with an increase in temperature. Emitter current stabilization can be realized by the use of external circuits that include temperature-sensitive devices. The circuit designer may employ several kinds of devices for this purpose. One such device is the *thermistor*. A thermistor has a negative temperature coefficient of resistance. That is, its resistance value decreases as the temperature increases. It is instructive to consider basic amplifier stage configurations that include thermistor control of emitter bias voltage.

Emitter Voltage Control Refer to Fig. 5–7. Here, an amplifier configuration is exemplified that employs a thermistor to vary the emitter voltage with temperature to minimize temperature variations in emitter current. This circuit comprises two voltage dividers, the first of which consists of R4 and R1, and the second of which consists of R2 and thermistor RT1. The first voltage divider permits the application of a portion of a battery voltage (V_C) to the base terminal and ground (common return). The base terminal voltage is developed across R1 and is in the *forward* bias polarity. The second voltage divider applies a portion of battery voltage V_C to the emitter terminal. This emitter terminal voltage is developed across R2 and is in the *reverse* bias polarity. Since the forward bias voltage applied to the base terminal is larger than the reverse bias applied to the emitter terminal, the resultant base-emitter bias is always in the forward direction.

With an increase in temperature, the collector current would normally increase when an unstabilized circuit is utilized. As noted previously, this increase in forward current can be prevented by reducing the forward bias. This is accomplished by the voltage divider comprising R2 and thermistor RT1. As the temperature increases, the resistance of RT1 decreases, causing more current to flow through the voltage divider. This increased current raises the negative potential at the emitter connection of R2. Thereby, the *reverse* bias applied to the emitter increases, and the net emitter-base forward bias decreases. Consequently, the collector current is reduced. If the temperature decreases, reverse actions take place to prevent the decrease of collector current. The circuit designer must analyze *worst-case conditions* on the basis of both thermistor and transistor tolerances as well as resistor tolerances.

Note that capacitor C1 blocks the DC voltage of the preceding stage and couples the AC signal into the base-emitter circuit. Capacitor C2 bypasses the AC signal around R2. Resistor R3 is the collector load resistor that develops the output signal. Capacitor C3 blocks the DC collector voltage from and couples the AC signal to the following stage.

Base Voltage Control A thermistor is employed to vary the base voltage with temperature in the circuit of Fig. 5-8 in order to minimize variations in emitter current with temperature. This circuit contains a voltage divider consisting of R1 and thermistor RT1. The voltage divider applies a portion of battery voltage V_C to the base-emitter circuit. Electron current flow through the voltage divider is in the direction of the arrow. This current produces a voltage of polarity indicated across RT1. A forward bias is applied to the transistor. If the temperature of the transistor increases, the emitter current would tend to rise. However, the resistance of thermistor RT1 de-

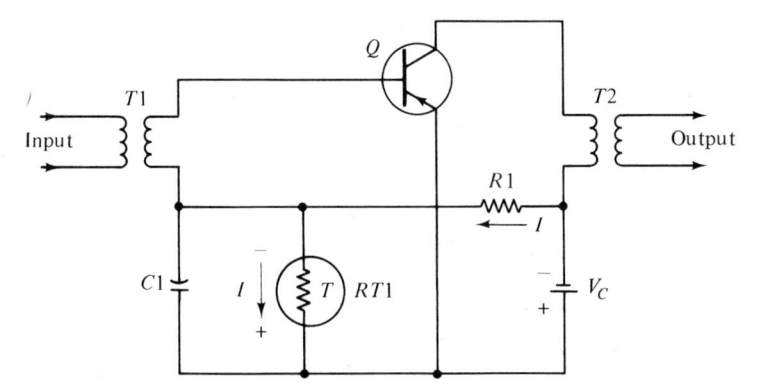

Figure 5-8 Amplifier stage configuration with thermistor control of base bias voltage.

creases with the increase in temperature, and causes more current flow through the voltage divider. This increased current flow causes a larger portion of the battery voltage V_C to drop across R1. In turn, the available voltage for forward bias, or the drop across RT1, is reduced, and the emitter current is thereby reduced.

Transformer T1 couples the AC signal into the base-emitter circuit. Capacitor C1 bypasses the AC signal around thermistor RT1. The primary winding of transformer T2 operates as the collector load, and develops the output signal, which in turn is coupled to the secondary winding on T2. As noted in the previous topic, *the circuit designer must make his worst-case analysis for this bias stabilization arrangement on the basis of tolerances on both Q and RT1, as well as resistor tolerances.*

Thermistor Limitations Thermistor capability to limit the variation of collector current with temperature in the amplifier arrangements of Figs. 5-7 and 5-8 is graphed in Fig. 5-9. One of the curves shows the variation in collector current for an unstabilized circuit. Another curve shows the variation in collector current for a thermistor-stabilized transistor. It is evident that there is appreciable improvement in bias stability when a thermistor is utilized. On the other hand, thermistor stabilization achieves *ideal stabilization* at only three temperatures (at *A, B,* and *C*). In other words, the thermistor resistance variation does not completely *track* the variation in emitter-base junction resistance. Therefore, the circuit designer may select some other bias-stabilization arrangement that provides improved tracking.

122 · Amplifier Bias Stabilization

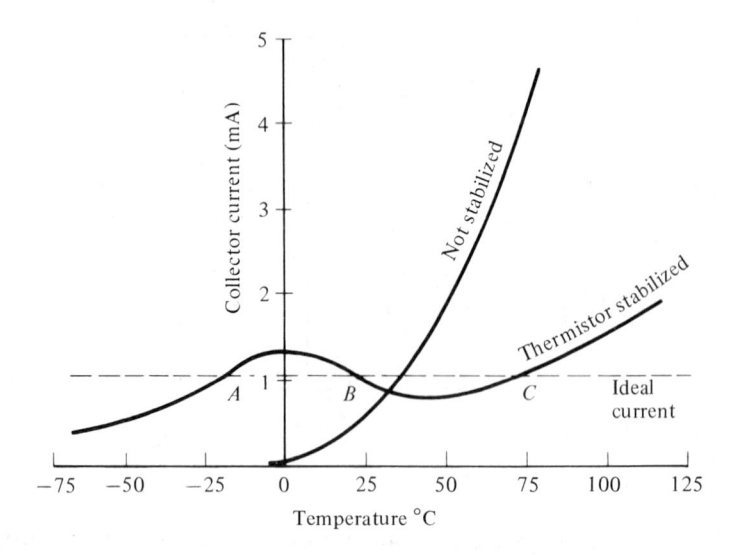

Figure 5-9 Variation of collector current with temperature for non-stabilized and thermistor-stabilized transistor amplifier stages.

5-5 DIODE STABILIZING CIRCUITS

Variation of collector current with temperature is caused by temperature variation of the emitter-base junction resistance and the saturation (reverse bias) current. Variation with temperature in the resistance and the reverse bias current of a PN junction occurs whether the PN junction is in a transistor or in a junction diode. A junction diode is a useful device in bias stabilization circuits for this reason. The chief advantage in employment of a junction diode as a temperature-sensitive device is that the circuit designer can choose a type of diode that has the same substances that are utilized in the associated transistor. The temperature coefficient of resistance for the diode and that of the transistor will then be the same. This approach permits a more constant collector current over a wide range of temperature because of better *tracking*.

Junction diodes have a negative temperature coefficient of resistance whether they are forward- or reverse-biased. It is instructive to consider bias stabilization means with a single forward-biased diode. The circuit depicted in Fig. 5-10 employs forward-biased junction diode CR1 as a temperature-sensitive device to compensate for variations of emitter-base junction resistance. Consider the voltage divider comprising R1 and junction diode CR1. The current I through the volt-

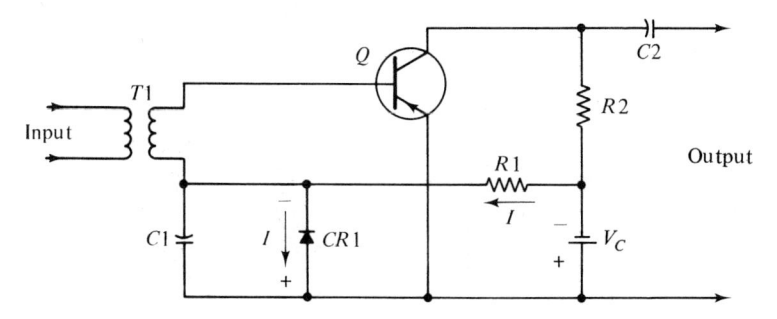

Figure 5-10 Transistor amplifier stage with a single forward-biased junction diode for compensation of emitter-base junction resistance with temperature.

age divider flows in the direction shown, and develops a voltage across CR1 with indicated polarity. This voltage provides a *forward* bias for Q. With increased temperature, the collector current would tend to increase. However, increased temperature decreases the resistance of CR1, causing more current to flow through the voltage divider. In turn, there is an increased voltage drop across R1. The voltage drop across CR1 is correspondingly decreased, thereby reducing the forward bias and the collector current through Q.

The AC signal is coupled into the transistor amplifier by transformer T1. Capacitor C1 bypasses the AC signal around diode CR1. Collector load resistor R2 develops the output signal. Capacitor C2 blocks the DC voltage from and couples the AC signal to the following stage. The effectiveness of this circuit to stabilize collector current with temperature is indicated by curve *BB* in Fig. 5-11. Compare this curve with curve *AA*, for which the transistor is not bias-stabilized, and also compare it with the ideal current level indicated in the diagram. Curve *BB* shows that a marked improvement in collector-current stability occurs for temperatures below 50°C. This indicates that the variation with temperature of the junction diode resistance tracks and compensates for the variations of emitter-junction resistance.

The sharp increase in collector current (curve *BB* in Fig. 5-11) at temperatures above 50°C indicates that junction diode CR1 *does not* compensate for the increase in saturation current (I_{CBO}). This condition might have been anticipated, since the saturation current (collector-base reverse bias current) flows out of the base, through T1 primary, through diode CR1, battery V_C, and back to the collector. Because the saturation current is a small percentage of the total

124 · Amplifier Bias Stabilization

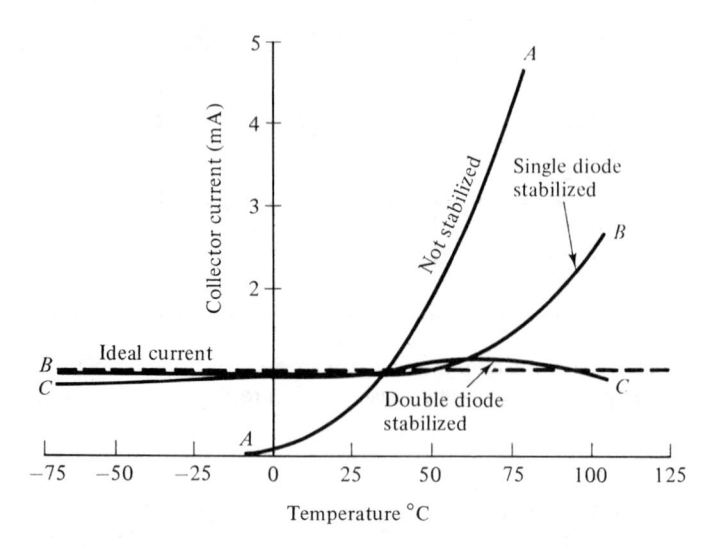

Figure 5–11 Graphs showing variation of collector current with temperature for non-stabilized, single-diode stabilized, and double-diode-stabilized transistor amplifier circuits.

current through diode CR1 (forward-biased and low in resistance), it causes no appreciable voltage drop across diode CR1. A circuit using a second junction diode (reverse-biased) is required to compensate for the saturation current, as explained next.

Double Diode Stabilization Two diodes are used as temperature-sensitive devices in the amplifier configuration of Fig. 5–12. One junction diode compensates for temperature variation in base-emitter junction resistance. The other diode compensates for temperature variation in saturation current. Resistor R3 and junction diode CR2 (reverse-biased) are included. Resistor R1 and diode CR1 (forward-biased) compensate for change in emitter-base junction resistance at temperatures below 50°C. Reverse-biased diode CR2 can be considered an open circuit at low temperatures. At room temperature, CR2 reverse-bias current I_s flows through CR2 in the direction indicated. Diode CR2 is *selected by the circuit designer* so that its reverse-bias current I_s is larger than that of the reverse-bias current I_{CBO} of the transistor. Diode reverse-bias current I_s consists of transistor reverse-bias current I_{CBO} and a component I_1 drawn from that battery. The voltage polarity developed by I_1 across R3 is indicated. Note that the emitter-base bias voltage is the sum of

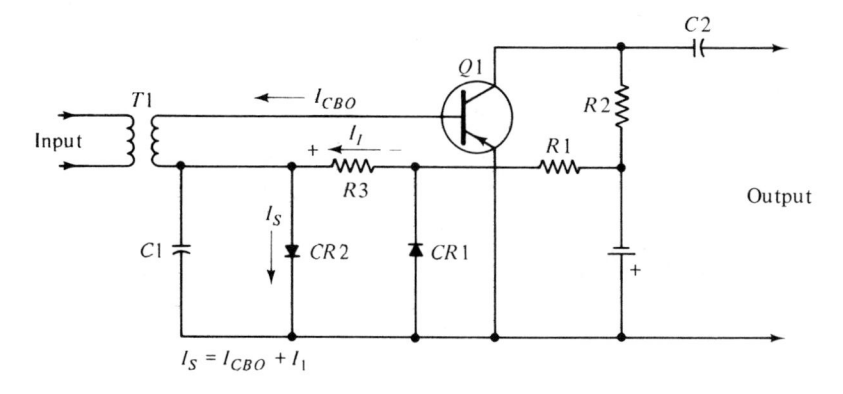

Figure 5–12 Amplifier configuration with double-diode stabilization against temperature variation.

the opposing voltages across R3 and CR1, if it is assumed that the secondary winding of T1 has negligible resistance.

As the temperature increases, currents I_{CBO}, I_s, and I_1 increase. The resultant *reverse-bias* voltage developed across R3 by current I_1 increases. The total *forward* bias (voltage across CR1 and R3) decreases with increasing temperature to stabilize the collector current. The functions of transformer T1, capacitors C1 and C2, and resistor R2 are the same as those for the corresponding elements of Fig. 5–10. In a *worst-case analysis* of this bias-stabilization method, the circuit designer must take into account the tolerances on both diodes, on the transistor, and on the resistors.

Reverse-biased Single-diode Stabilization A reverse-biased junction diode CR1 is used in the circuit of Fig. 5–13 as a temperature-sensitive device. This amplifier configuration provides two separate paths for the two components of the base current. The base-emitter current $(I_e - \alpha_{fb})$ flows through the base region to the emitter, through R2, battery V_C, and R1 to the base lead. The saturation current I_{CBO} flows from the base lead through CR1, battery V_B, R3, and the collector region to the base region. *The circuit designer selects the diode* so that its saturation (reverse bias) current equals that of the transistor over a wide temperature range.

As the temperature increases, I_{CBO} increases. However, the saturation of CR1 increases by an equal amount, so that there is no accumulation of I_{CBO} current carriers in the base region. Such an accumulation would cause an increase in emitter current. Diode CR1

126 · Amplifier Bias Stabilization

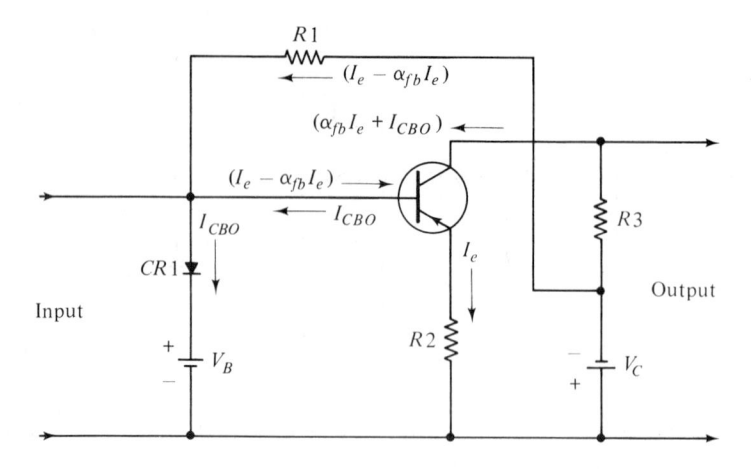

Figure 5–13 Transistor amplifier configuration using a single reverse-biased diode for compensation of temperature variations in saturation current.

acts as a gate, and opens wider to accommodate the increase in I_{CBO} with temperature. This circuit is utilized by the designer if *RC* coupling is used between it and the previous stage. The reverse-biased diode provides high resistance for the input circuit. Resistor R2 swamps the emitter-base junction resistance, and prevents a large increase in emitter current, particularly at low temperatures. Resistor R3 is the collector load resistor and develops the AC output signal.

5–6 TRANSISTOR STABILIZED AMPLIFIER CIRCUITS

Certain currents and voltages of a temperature-stabilized transistor may be used to temperature-stabilize another transistor or several transistors. Consider the utilization of a common emitter-base voltage for this purpose. The emitter-base junction of a transistor has a negative temperature coefficient of resistance similar to that of a PN junction diode. *It is possible to use the variations of the emitter-base junction resistance of one transistor to control the emitter-base bias of a second transistor.* A circuit that employs this principle is exemplified in Fig. 5–14. The emitter-base voltage of Q1 is used to bias the emitter-base junction of transistor Q2. If zero DC resistance is assumed in the secondary winding of T2, the base of Q2 has a direct DC connection to the emitter of Q1. If zero DC resistance is assumed in the secondary of T1, the emitter of Q2 has a direct

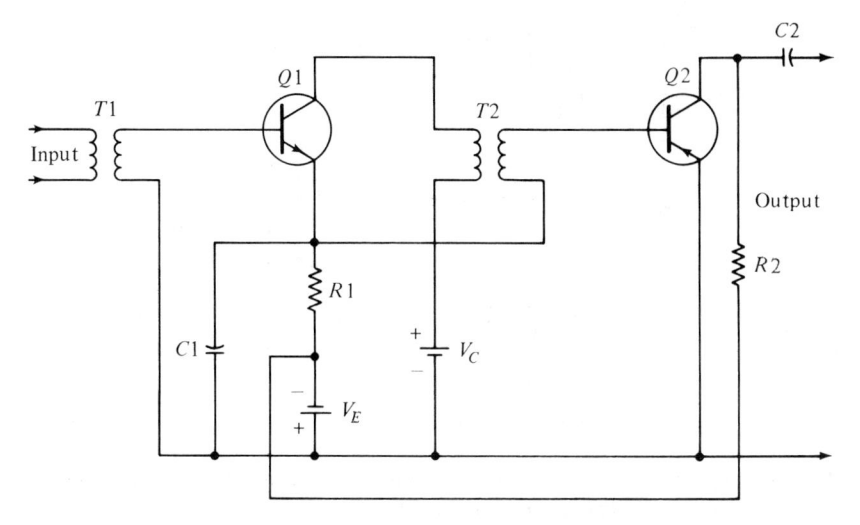

Figure 5-14 An amplifier configuration in which temperature stabilization of Q2 is provided by the emitter-base voltage of Q1.

DC connection to the base of Q1. Battery V_E provides forward bias for Q1. Cross-connection of the emitters and bases of the two transistors provides forward bias for Q2, because it is a PNP type.

Transistor Q1 is temperature-stabilized by the use of a high-valued resistor R1 (swamping resistor) in the emitter lead, and low DC resistance in the base lead. Resistor R1 maintains a relatively constant emitter current in Q1. As the temperature increases, the base-emitter junction resistance of Q1 decreases. Because the current through the junction remains constant, the voltage drop across the junction decreases. A decrease in this voltage represents a decrease in the forward bias on Q2. In turn, the tendency of the collector current through Q2 to increase with temperature is offset. If a curve of collector current through Q2 were plotted versus temperature, it would be similar to the curve *CC* shown in Fig. 5-11.

In the amplifier circuit of Fig. 5-14, transformer T1 couples the AC signal input to the base-emitter circuit of Q1. Battery V_E supplies base-emitter bias voltage for Q1 and collector voltage for Q2. Resistor R1 operates as an emitter swamping resistor. Capacitor C1 bypasses AC signal around R1 and battery V_E. Battery V_C supplies Q1 collector voltage. Transformer T2 couples the AC signal output of Q1 to the input circuit of Q2. Resistor R2 is the collector load resistor for Q2 and develops the output signal. Capacitor C2

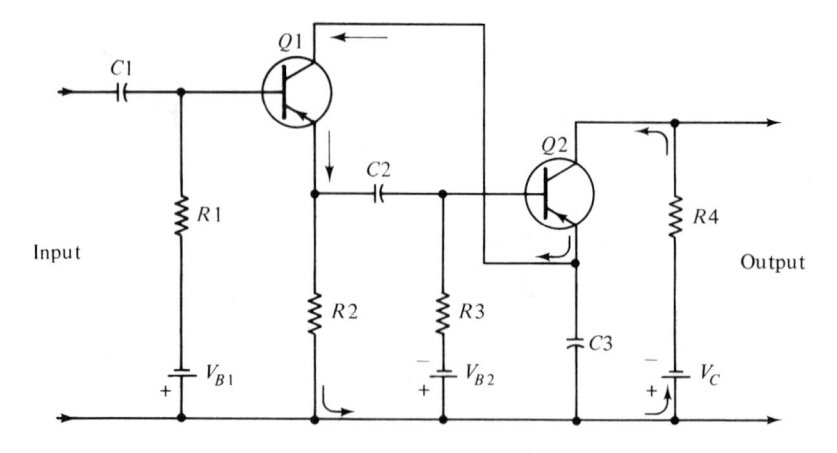

Figure 5–15 Amplifier configuration with common emitter-collector current for temperature stabilization of two transistors.

blocks the DC voltage from and couples the AC signal to the following stage.

Common Emitter-collector Circuit A method of temperature-stabilizing the emitter-collector current of one transistor by using the stabilized emitter-collector current of another transistor is shown in Fig. 5–15. Transistor Q1 collector current is stabilized by use of swamping resistor R2. The circuit designer selects R1 so that its value is low, thus limiting accumulation of saturation current carriers in the base region. The stabilized DC collector current of Q1 is made to flow through the emitter-collector region of Q2 by connecting the collector of Q1 directly to the emitter of Q2. The current direction of this circuit is indicated by the arrows. Electrons that flow into the emitter lead of Q1 flow through R2, battery V_C, R4, collector-emitter of Q2, and into the collector of Q1.

The advantage of this method of stabilizing the collector current of Q2 is that it eliminates the need for a swamping resistor in the emitter lead of Q2. *The elimination of such a swamping resistor is particularly important in power amplifiers that draw heavy emitter currents.* The functions of the circuit elements are as follows: Note that Q1 is used in a CC configuration, and that Q2 is used in a CE configuration. Capacitor C3 places the collector of Q1 and the emitter of Q2 at AC zero potential. Capacitor C1 blocks the DC voltage from the previous stage and couples the AC signal to the base of

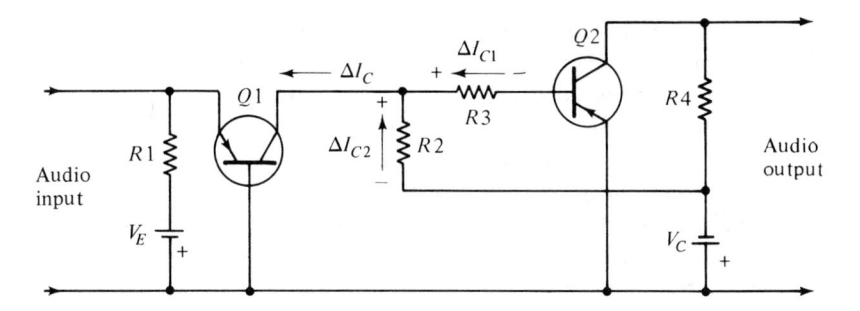

Figure 5-16 Two-stage direct-coupled amplifier configuration with collector-current temperature stabilization.

Q1. Resistor R1 provides a DC return path for base current. Battery V_{B1} provides base-emitter bias. Resistor R2 is the emitter load resistor and develops the output signal. Capacitor C2 blocks the emitter DC voltage from and couples the AC signal to the base of Q2. Resistor R3 provides a DC return path for base current. Battery V_{B2} provides base bias. Resistor R4 is the collector load resistor and develops the output signal. Battery V_C provides voltage for both transistors.

Bias Variation with Collector Current Two stages of direct-coupled amplification are depicted in Fig. 5-16. A DC amplifier amplifies a DC input voltage and very low frequency AC signals in addition to the high frequency range that is accommodated by a corresponding *RC*-coupled amplifier. This circuit is designed so that an increase in collector current caused by a temperature rise in Q1 will reduce the forward bias in Q2. The circuit operates as follows: Transistor Q1 operates in the CB mode; its stability factor is ideal. However, some variation of its collector current occurs with change in temperature. Thus, assume that an increment in collector current (ΔI_C) occurs. This increase in collector current is in the direction of the arrow. A portion of the increment in collector current flows through R3. This current is designated ΔI_{C1} and develops a voltage across R3 with the indicated polarity.

Another portion of the increment in collector current flows through R2. This current is designated ΔI_{C2} and develops a voltage across R2 with the indicated polarity. Note that the voltage polarities indicated are only for an *increment* in collector current, and are not necessarily the steady-state voltage polarities. The steady-state voltages are of no concern in the bias-stabilization analysis. Consider the base-emitter bias circuit of Q2. The base-emitter bias is the sum of the voltages across R3, R2, and battery V_C. The indicated voltage across R3 *aids* the forward bias; that across R2 *opposes* the forward bias. *The circuit designer selects the values of R2 and R3 so that the voltage across R2 is the larger; thereby, the resultant forward bias is decreased.* This action limits the tendency of the Q2 collector current to increase with temperature.

R1 is the emitter swamping resistor. Battery V_E supplies emitter bias voltage. Resistor R4 is the collector load resistor. The voltage stability factor is expressed by an equation noted previously; an inspection of this equation shows that the voltage stability factor is directly proportional to the current stability factor (an equation also noted previously). Therefore, *the circuit techniques for improving the current stability factor will also improve the voltage stability factor.* In other words, the variations of collector voltage with temperature are limited as a consequence of temperature variations in saturation current.

5–7 BREAKDOWN DIODE VOLTAGE REGULATOR

A graph of the current through and the voltage across a reverse-biased diode junction is exemplified in Fig. 5–17(a). Observe that at a certain value of reverse-bias voltage E_{out}, the current increase is rapid, whereas the voltage drop across the diode remains essentially constant. This is called zener diode action, and it has useful applications in bias stabilization methods. The voltage-regulator circuit shown in Fig. 5–17(b) is capable of maintaining constant load voltage, although variation may occur in the source voltage or in the load current. Depending upon the characteristics of the particular zener diode, the breakdown voltage can be any value from 2 to 60 volts.

Breakdown Diode Temperature Compensation A reverse-biased diode junction has a negative temperature coefficient of resistance. This characteristic is limited to a reverse-bias voltage that does not equal or exceed the breakdown voltage. A zener diode has a *positive* temperature coefficient of resistance several times larger than the *negative* temperature coefficient of resistance for a forward-biased or reverse-biased junction diode. Note that the discussion in the previous topic concerning the voltage regulator in Fig. 5–17(b) ap-

Breakdown Diode Voltage Regulator · 131

(a)

(b)

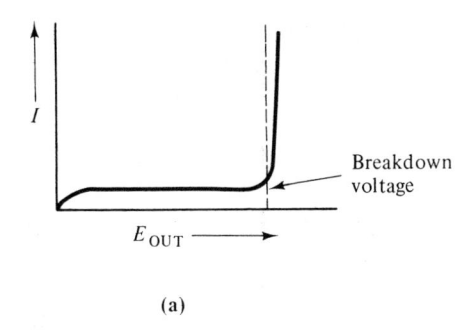

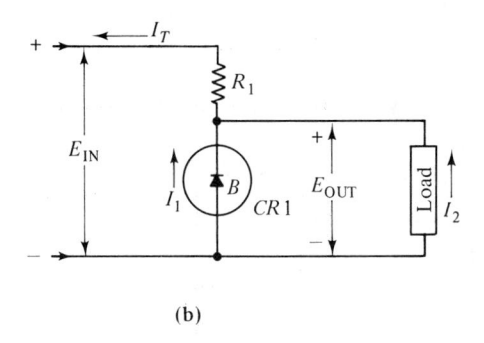

Figure 5–17 Breakdown diode voltage-current characteristic, and voltage-regulator circuit.

plies only if the temperature of the zener diode does not vary under operating conditions. One method of compensating for the increasing or decreasing breakdown-diode resistance with increasing or decreasing temperature, respectively, is to place elements of negative temperature coefficient in series with the breakdown diode. In Fig. 5–18, a circuit is shown with two forward-biased diodes (CR1 and CR2) in series with breakdown diode CR3. The total resistance of the three diodes in series remains constant over a wide range of temperature. The net result is a constant-voltage output, although temperature, input voltage E_{IN}, and load-current drain may vary. Two diodes are used in this circuit, because the temperature coefficient of resistance of each is half that of the breakdown diode. Forward-biased diodes are utilized because of the very low voltage drop across them. Note that *the circuit designer could use thermistors, or other temperature-sensitive devices, if he chose.*

132 · Amplifier Bias Stabilization

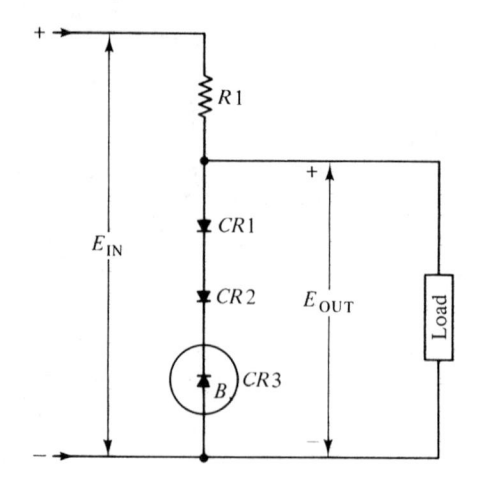

Figure 5-18 Breakdown diode temperature-compensated voltage regulator.

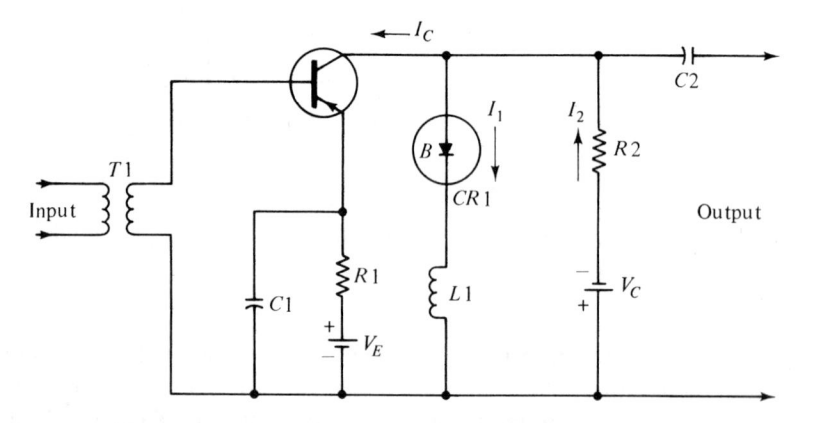

Figure 5-19 Breakdown diode used in an amplifier circuit to provide collector voltage stabilization.

Voltage Stabilized Transistor Amplifier The circuit shown in Fig. 5-19 uses a breakdown diode (CR1) to stabilize the collector voltage. Note that the current I_2 that is drawn from the battery divides into the breakdown diode current I_1, and into the collector current I_C. When the DC collector current increases because of an increase in temperature, current I_1 decreases by the same amount, so that current I_2 remains constant. The DC voltage drop across R2 also remains constant. The voltage applied to the collector (the battery voltage less the voltage drop across R2) also remains constant. The foregoing discussion neglects the variation in resistance with temperature of the breakdown diode CR1. To obviate this factor, two additional diodes can be included, as depicted in Fig. 5-18.

The AC resistance of a breakdown diode may vary from 5 ohms to 1000 ohms, depending upon device design. To avoid shunting collector load resistor R2 by the low AC resistance of diode CR1, a high impedance choke coil L1 is connected in series with the diode. Transformer T1 couples the AC signal to the base-emitter circuit. Resistor R1 is the emitter swamping resistor. Battery V_E provides emitter-base bias voltage. Capacitor C1 bypasses the AC signal around resistor R1 and battery V_E. Capacitor C2 blocks DC voltage from and couples the AC signal to the following stage.

Surge Protection by Junction Diode If an excessive emitter-collector voltage occurs when the normally forward-biased base-emitter circuit is *reverse-biased,* internal oscillation may occur that can destroy the transistor. This hazard may exist in transistor amplifiers that employ transformer input and output, as exemplified in Fig. 5-20. If the signal input from the previous stage is suddenly terminated, or if excessively strong noise signals drive the base-emitter circuit into a reverse-bias condition, the collector current cuts off rapidly. In turn, the magnetic field surrounding the windings in T2 collapses rapidly and generates a high emitter-collector counter emf and reverse-biases the base-emitter junction. To forestall possible destruction of the transistor, *the circuit designer may connect a junction diode between the base and emitter terminals to prevent the base-emitter junction from becoming reverse-biased.*

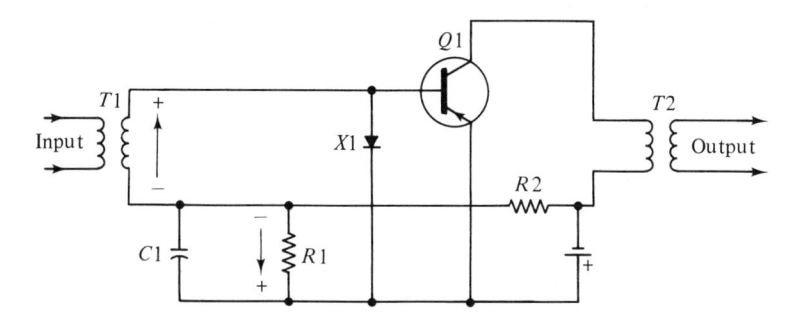

Figure 5-20 Junction diode action prevents the base-emitter junction of the transistor from becoming reverse-biased.

Note that the voltage divider comprising R1 and R2 supplies a voltage into the base circuit that forward-biases the base-emitter junction, and that reverse-biases the junction diode (CR1). Under normal operating conditions, CR1 is effectively an open circuit. Consider a strong voltage surge of the indicated polarity to occur across the secondary of T1. If the surge voltage exceeds the voltage across R1, CR1 becomes forward-biased and conducts. In its conducting state, only a very small voltage is dropped across CR1. This drop is negligible and prevents the base-emitter junction of the transistor from becoming reverse-biased.

5-8 INNOVATIVE DESIGN

Innovative design denotes a departure from established forms, or an introduction of new methods. An innovative approach may be taken to reduce production costs, to avoid patent infringement, or to enhance the market appeal of a product. As an illustration, a conventional power transformer has flexible leads to be cut, stripped, and individually connected to a circuit board. The engineer who designed power transformers with lug terminals that could be simultaneously inserted into a circuit board achieved an innovative design that provided a significant reduction in production costs. Again, the engineer who designed a one-piece control knob-and-shaft unit achieved an important innovative design. Although modular construction is a commonplace today, it represented an eventful example of innovative design when it was originally introduced.

Today, the sophisticated circuit designer can find a fruitful field for innovation in various forms of digital systems simulation. Digital logic simulation is used for logic verification, design verification, or both. The potential for achievement in this area of circuit design is limited only by the ability and ingenuity of the engineer.

CHAPTER 6

Fundamentals of Radio-Frequency Circuit Design

6–1 GENERAL CONSIDERATIONS

Most radio-frequency circuits operate in a series-resonant or in a parallel-resonant mode. Aside from the practical necessity for selectivity in many applications, the *efficiency* of a radio-frequency (RF) amplifier is greatly increased by operating its input circuit and its output circuit in a resonant condition. Efficiency is enhanced because any active device has junction capacitances, and all practical circuit arrangements have stray capacitance. Although these capacitances can be neglected at audio frequencies, they become a dominant consideration at radio frequencies because capacitive reactance is inversely proportional to the operating frequency. Efficiency is also enhanced in various design situations, because tapped resonant circuits can be employed to match the output impedance of one device to the input impedance of a following device. In other words, maximum power is transferred from a source to a sink when their impedances are matched; although system efficiency is only 50 percent at maximum power transfer (Fig. 6–1), this trade-off is acceptable in the vast majority of designs.

Resonant circuits avoid the bypassing action of stray capacitance and junction capacitance by utilizing these capacitances in tuning an inductor to resonance at the operating frequency. At this point, it is instructive to consider *the effect of tolerances on source and load impedances with respect to maximum power transfer.* An inspection of the power variation in Fig. 6–1(a) shows that this is

135

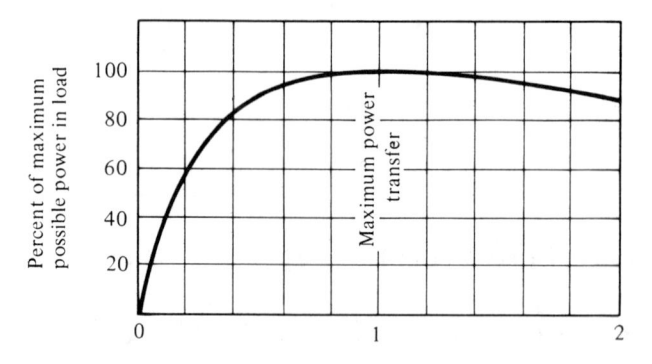

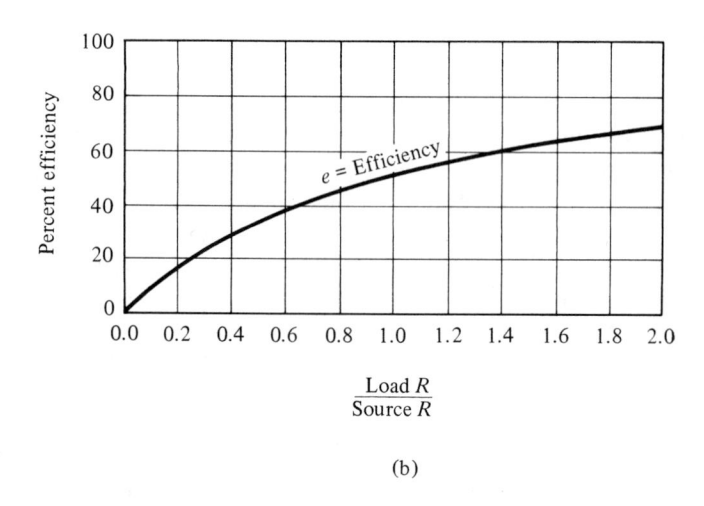

136 · Fundamentals of Radio-Frequency Circuit Design

Figure 6–1 Power transfer and efficiency relations for source and load: **(a)** power transfer versus impedance ratio; **(b)** system efficiency impedance ratio.

not a highly critical design consideration. In other words, a mismatch of ±40 percent reduces the power transfer less than 10 percent. On the other hand, if the load is low and departs 80 percent from optimum, 40 percent of the available power is lost. If the load value is high and departs 80 percent from optimum, the loss is much less and over 90 percent of the available power is still delivered to the load.

Basic Resonant Circuit Design   137

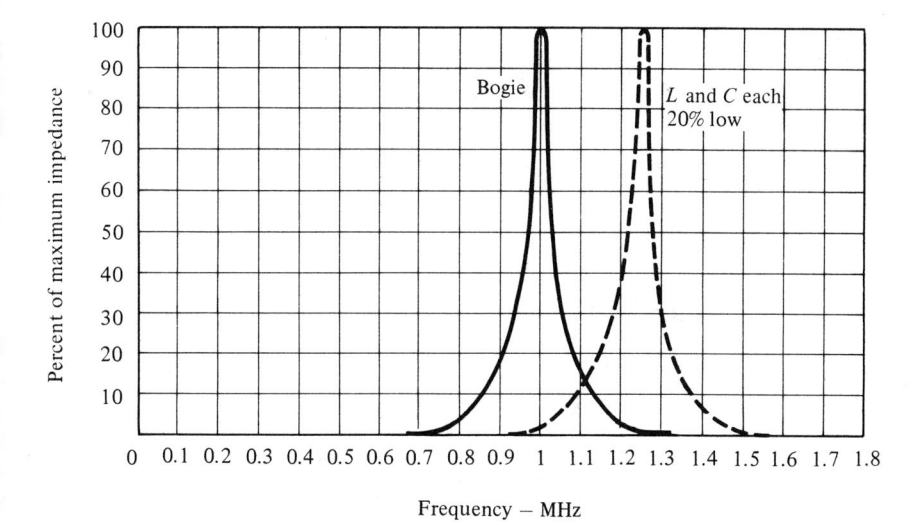

Figure 6–2 Error in resonant frequency, owing to 20 percent tolerances on L and C.

6–2 BASIC RESONANT CIRCUIT DESIGN

With reference to the four forms of parallel-resonant circuits shown in Fig. 6–3, the basic resonant-frequency equation is

$$f_r = \frac{1}{2\pi\sqrt{LC}}$$

This equation is exact for a resistanceless LC parallel configuration, and it is a good approximation for an LCR parallel-resonant configuration. Consider the effect of component tolerances on the resonant-frequency value. If the tolerance on L and on C is ±20 percent, the worst-case condition occurs when both components have their low-limit values. In this case, the resonant frequency shifts 25 percent above its bogie value, as shown in Fig. 6–2. The *quality factor Q* of a parallel-resonant circuit is equal to the ratio of tank current I_L or I_C to the line current I, as depicted in Fig. 6–3. This equation is *not* true for operation off-resonance.

The circuit designer should note that *if the Q value of a parallel-resonant circuit is high, the line current I will be small, whereas the capacitor current I_C and the inductor current I_L will be large.* In turn, normal operation can be obtained only if the designer provides

138 · Fundamentals of Radio-Frequency Circuit Design

$X_L = 2\pi fL$

$X_C = \dfrac{1}{2\pi fC}$

$Q = \dfrac{I_C}{I} = \dfrac{X_L}{R} = \dfrac{Z}{X_C}$

(a)

(b)

(c)

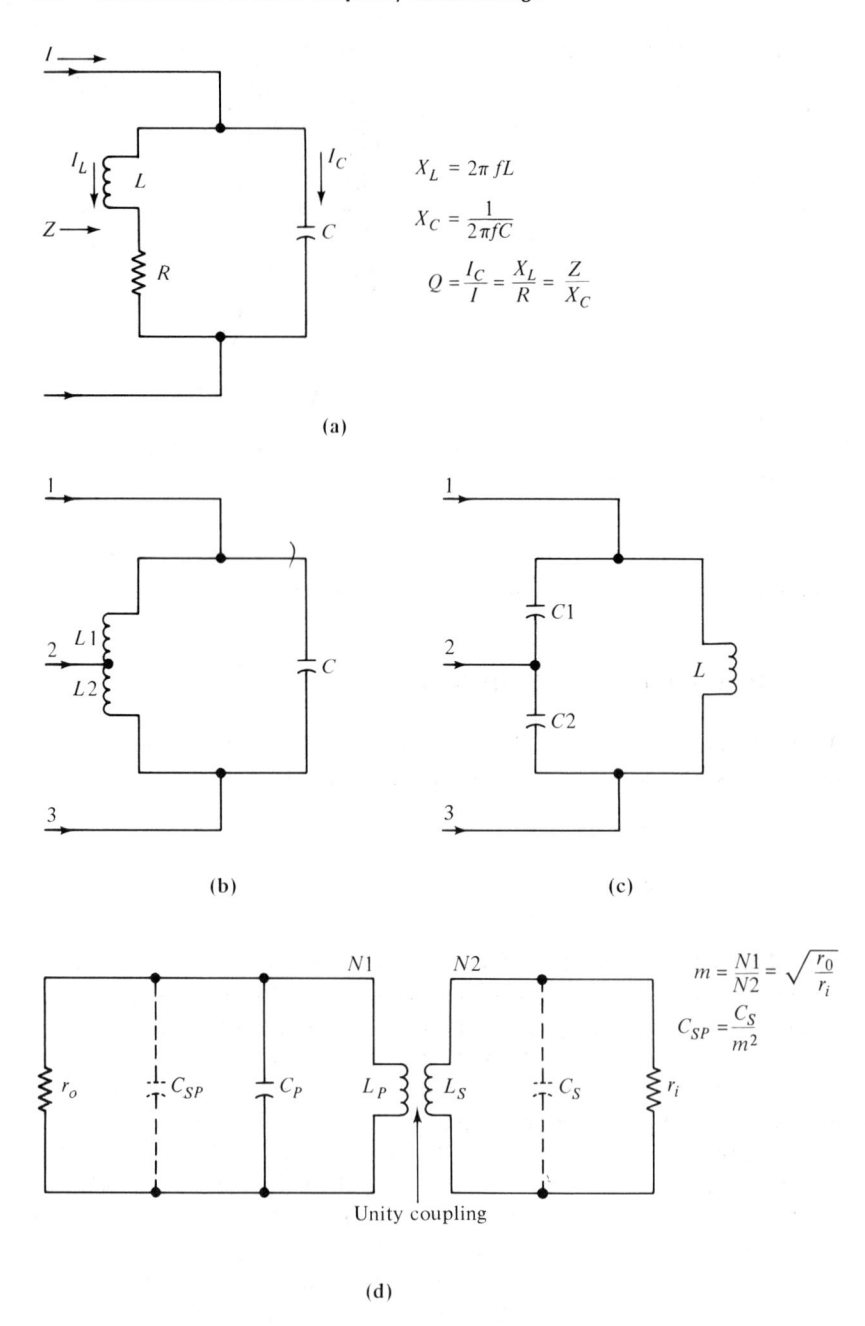

$m = \dfrac{N1}{N2} = \sqrt{\dfrac{r_0}{r_i}}$

$C_{SP} = \dfrac{C_S}{m^2}$

Unity coupling

(d)

Figure 6–3 Fundamental relations in basic parallel-resonant circuits.

an inductor that can handle the large RF current without undue loss. Similarly, the inductor must be connected to the capacitor with leads that have low RF resistance. In most cases, the capacitor is a lesser problem; most capacitors have a low dissipation factor and can handle comparatively large RF currents satisfactorily. Note that the Q value of a component is equal to its reactance (at a specified frequency), divided by its RF resistance. The dissipation factor D of a component is equal to the reciprocal of its Q value.

To measure the RF resistance of an inductor at a specified operating frequency, the designer measures its Q value and its inductance value with an LCR bridge. In turn, its RF resistance is equal to its reactance at the specified frequency divided by its Q value. This RF resistance value will always be greater than the DC resistance value of the inductor, and may be very much larger at high frequencies. Skin effect, proximity effect, possible eddy currents, dielectric losses, and radiation loss all contribute to the RF resistance value. The Q value of a parallel-resonant circuit is a measure of its *selectivity*. In other words, the resonant circuit has a bandpass value equal to the number of hertz between the points on its resonance curve that are 0.707 (−3 dB) down from the peak of the curve. An approximate equation states:

$$BW \approx \frac{f_r}{Q}$$

where BW is the resonant-circuit bandwidth in hertz.
f_r is the resonant frequency.
Q is the quality factor of the inductor.

Note that the Q factor of the parallel-resonant circuit would be reduced if the RF losses in the capacitor were significant. However, as noted previously, the designer can ordinarily neglect capacitor losses. *Errors in design bandwidth* result from tolerances on the RF resistance of the inductor. The *worst-case condition* occurs when the RF resistance is at its lower tolerance limit. For example, if the RF resistance is 20 percent low, the Q value will be 25 percent above bogie. In turn, the bandwidth will be 20 percent less than bogie.

Tapped inductors are widely used in solid-state circuit design. With reference to Fig. 6–3(b), the resonant frequency and the unloaded Q value of a parallel-resonant circuit remain the same whether the signal is injected across points 1 and 3, or across points 2 and 3. *The input impedance between points 2 and 3, however, is*

140 · Fundamentals of Radio-Frequency Circuit Design

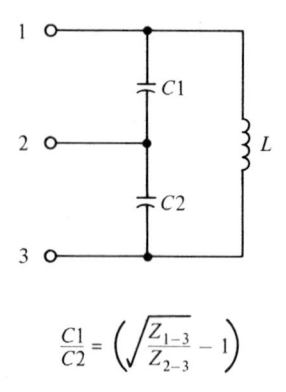

$$\frac{C1}{C2} = \left(\sqrt{\frac{Z_{1-3}}{Z_{2-3}}} - 1\right)$$

Figure 6–4 Relation of impedance ratios to capacitance ratios.

much smaller than the input impedance between points 1 and 3, depending on the square of the turns ratio between points 1 and 3, and between points 2 and 3. For example, if point 2 is a center-tap, the input impedance between points 1 and 3 is *four times* the input impedance between points 2 and 3. In other words, if point 2 were one-third up from point 3, the input impedance between points 1 and 3 would be nine times the input impedance between points 2 and 3. To continue the foregoing example, the input impedance between points 1 and 3 would be 2-1/4 times the input impedance between points 2 and 3. *An error of 20 percent in center-tap location on a coil results in a 56 percent worst-case impedance error.*

Consider next the tapped arrangement depicted in Fig. 6–3(c). The resonant frequency and the unloaded Q remain the same, whether the signal is injected across points 1 and 3, or across points 2 and 3. However, the input impedance between points 2 and 3 is smaller than the input impedance between points 1 and 3. For example, if C1 and C2 have equal values, the input impedance between points 1 and 3 is *four times* the input impedance between points 2 and 3. This fact is evident, because if point 2 were connected to a center tap on L, circuit action would remain unchanged inasmuch as a balanced bridge configuration results. The general C versus Z relation is shown in Fig. 6–4.

In many design situations, a tuned resonant circuit (capacitor C_p and inductor L_p) in the primary of a transformer is coupled to the non-resonant secondary of the transformer, as depicted in Fig. 6–3(d). In this case, if $N1$ represents the number of turns in the primary winding, and $N2$ represents the number of turns in the secondary winding, then the turns ratio **m** of primary to secondary under matched conditions is

$$m = \frac{N1}{N2} = \sqrt{\frac{r_o}{r_i}}$$

where r_o and r_i are the impedances (resistances) to be matched.

If there is capacitance in the secondary circuit C_S, it is reflected (or referred) to the primary circuit by transformer action as a capacitance C_{SP}. Its effective value in the primary circuit is

$$C_{SP} = \frac{C_S}{m^2}$$

It is evident that the worst-case condition for the value of **m** occurs when $N1$ has its high-tolerance limit and $N2$ has its low-tolerance limit. For example, *a 10 percent winding tolerance under these conditions corresponds to a 22 percent error in the **m** value.*

6–3 TRANSISTOR AND COUPLING NETWORK IMPEDANCES

The output impedance of a transistor can be considered as a resistance r_o in parallel with a capacitance C_o, as shown in Fig. 6–5(a). Similarly, the input impedance of a transistor can be considered as a resistance r_i in parallel with a capacitance C_i, as indicated in the diagram. In most situations, the circuit designer accounts for the output capacitance C_o and the input capacitance C_i by considering them to be part of the coupling network, as represented in Fig. 6–5(b). Note in passing that *there is an appreciable tolerance on the junction capacitances of a transistor,* which can seldom be ignored in design procedures. Assume that the required capacitance between terminals 1 and 2 of the coupling network is calculated to be 500 pF. Assume also that capacitance C_o is 10 pF. A capacitor with a value of 490 pF would then be utilized between terminals 1 and 2 so that the total capacitance would be 500 pF. The same method is used to compensate for capacitance C_i between terminals 3 and 4.

To cope with a production tolerance of ±20 percent on the value of C_o, the circuit designer could use a capacitor with a value of 488 pF, paralleled by a trimmer capacitor with a maximum value of 4 pF. Then the trimmer capacitor would be adjusted in production test procedure to resonate the coupling network correctly. To

142 · Fundamentals of Radio-Frequency Circuit Design

Transistor $Q1$ output   Coupling network   Transistor $Q2$ input

(a)

Transistor $Q1$ output   Coupling network   Transistor $Q2$ input

(b)

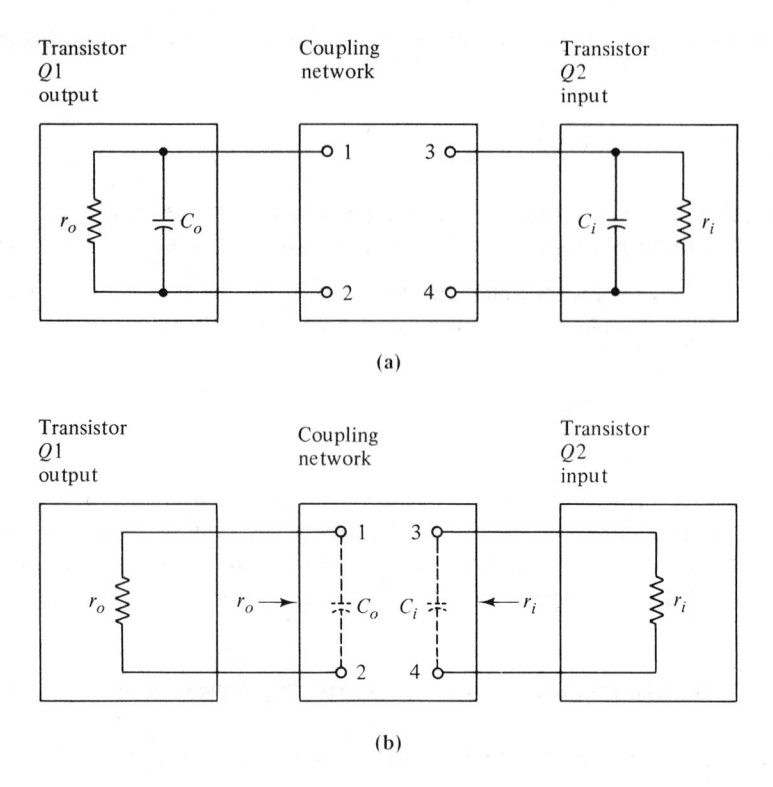

Figure 6–5 Equivalent output and input circuits for transistors with a coupling network.

obtain maximum power transfer from Q1 to Q2, the input impedance (terminals 1 and 2) to the coupling network must equal resistance r_o; the output impedance of the coupling network (*looking* into terminals 3 and 4) must equal r_i.

6–4 TRANSFORMER COUPLING WITH TUNED PRIMARY

A generalized representation of a single-tuned circuit used as a coupling network, and utilizing inductive coupling, is shown in Fig. 6–6(a). In this circuit, capacitance C_T represents the output capacitance of transistor Q1 and the input capacitance of transistor Q2 referred to the primary of transformer T1. A match of the output impedance of Q1 to the input impedance of Q2 is obtained by means of a suitable turns ratio for T1. If L_P represents the inductive reactance between terminals 1

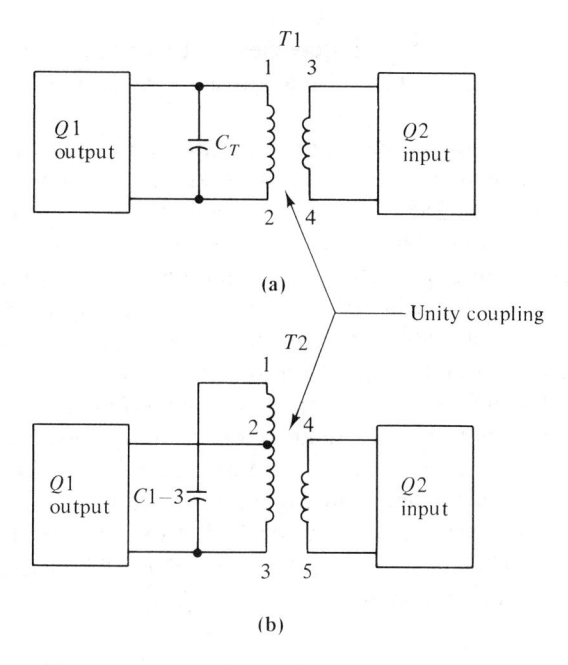

Figure 6–6 Interstage network utilizing transformer coupling with tuned primary winding.

and 2 of T1, then the resonant frequency f_r for this circuit is given by the equation:

$$f_r = \frac{1}{2\pi\sqrt{L_P C_T}}$$

Suppose that the circuit designer employs the configuration in Fig. 6–6(a) for an intermediate-frequency (IF) amplifier with two transistors operated in the CE mode. In turn, the basic design parameters are

Resonant frequency $f_r = 500$ kHz.
Frequency bandwidth $\Delta f = 10$ kHz.
Transistor Q1 output resistance $r_o = 12{,}000$ ohms.
Transistor Q1 output capacitance $C_o = 10$ pF.
Transistor Q2 input resistance $r_i = 700$ ohms.
Transistor Q2 input capacitance $C_i = 170$ pF.
Coupling network insertion loss = 3 dB.

Power response of frequencies at extremes of bandwidth ($f_r \pm \Delta f/2$) = 1/4 of response at resonant frequency f_r.

With these data and requirements, it can be shown that the unloaded Q value of the tuned circuit (before it is connected into the amplifier) must be 288. The primary inductance L_P must be 7.7 μH, the secondary inductance L_S must be 0.45 μH, and capacitor C_T must have a value of 13,100 pF (including C_o = 10 pF and C_i = 10 pF referred to the primary). To construct a transformer with the high Q value of 288 and with the low inductance value of L_P = 7.7 μH is very difficult. *Therefore, the circuit designer should circumvent the difficulty by using a tapped primary winding,* as depicted in Fig. 6–6(b). The inductance of the primary can be made many times the calculated value. For example, the inductance $L_{1\text{-}3}$ between terminals 1 and 3 in Figure 6–6(b) can be made 100 times the previously calculated value for L_P between terminals 1 and 2 in Fig. 6–6(a). That is, $L_{1\text{-}3}$ may have a value of 770 μH, and it becomes practical to construct a tuned circuit with an adequate Q value.

In turn, to maintain the same resonant frequency, the capacitance connected across terminals 1 and 3 ($C_{1\text{-}3}$) must be reduced to 0.01 of the previously calculated value for C_T. Thus, a capacitance of 131 pF is connected across terminals 1 and 3. Next, to maintain a matched impedance condition for maximum power transfer, the inductance between terminals 2 and 3 of transformer T2 must be equal to the previously calculated value of L_P between terminals 1 and 2 of transformer T1, or 7.7 μH. *To contend realistically with production tolerances on components and devices, the circuit designer may provide a ferromagnetic tuning slug in L2.* Note that a tuning slug can improve the Q value of a coil.

Autotransformer Coupling, Tuned Primary Inductive coupling from the output of one transistor to the input of a second transistor can be accomplished with an autotransformer, as shown in Fig. 6–7(a). The circuit action is essentially the same as in the configuration of Fig. 6–6, except that DC isolation is not provided between transistors Q1 and Q2. Capacitance C_T in Fig. 6–7(a) includes the output capacitance of Q1 and the input capacitance of Q2 referred to the primary section. If $L_{1\text{-}3}$ designates the inductance between terminals 1 and 3 of the autotransformer, then the resonant frequency of the circuit is expressed:

$$f_r = \frac{1}{2\pi \sqrt{L_{1\text{-}3} C_T}}$$

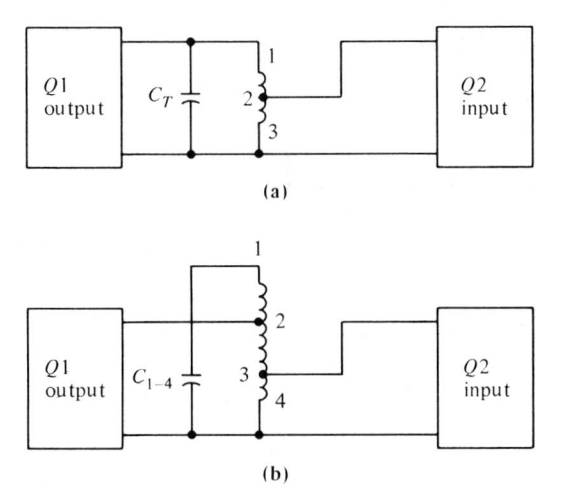

Figure 6-7 Interstage network utilizing autotransformer coupling with a tuned primary winding.

The tap at terminal 2 on the autotransformer is specified to provide an impedance match between Q1 and Q2, as explained previously. In some applications, a problem may arise owing to poor selectivity; when this occurs, the Q value between terminals 1 and 3 is too low. To increase the Q value, the primary inductance should be increased. *The circuit designer can circumvent this difficulty by utilizing the alternative autotransformer arrangement depicted in Fig. 6-7(b).* To maintain the same resonant frequency, C_{1-4} must be reduced by the same factor so that the product of L_{1-3} and C_T in Fig. 6-7(a) is equal to the product of L_{1-4} and C_{1-4} in Fig. 6-7(b). Impedance matching is maintained, provided that inductances L_{2-4} and L_{3-4} of transformer T2 equal the inductances L_{1-3} and L_{2-3}, respectively, of transformer T1.

Capacitance Coupling When the designer encounters a transformer coupling problem that limits the secondary to a small number of turns, it may be impractical to realize unity coupling, or even tight coupling, between primary and secondary. This problem is particularly troublesome in CB configurations that operate at very high frequencies. In other words, the input impedance to the transistor is less than 75 ohms in this situation. *To circumvent this difficulty, the circuit designer may utilize capacitive coupling,* as exemplified in Fig. 6-8. Impedance matching of Q1's output impedance to Q2's in-

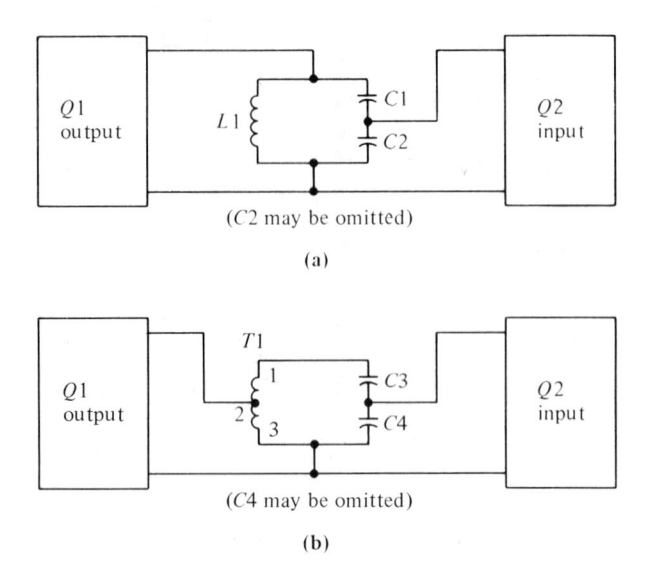

Figure 6-8 Interstage capacitance coupling with a split capacitor.

put impedance is obtained by selecting the proper ratio of C_1 to C_2. Capacitance C_2 is ordinarily much larger than C_1. Their reactance values bear the opposite relationship to their capacitance values. In the CB configuration, particularly, C_2 may be omitted because its reactance is masked by the low input resistance.

If the designer finds that inductance L_1 is too small to provide the required selectivity, he may utilize the configuration shown in Fig. 6-8(b). With the coil tapped, the inductance may be increased by a chosen factor, and the total capacitance C_T reduced by the same factor in order to maintain the same resonant frequency. Impedance matching is realized by making the inductance between terminals 2 and 3 equal to L_1; the ratio of C_3 to C_4 must be equal to the ratio of C_1 to C_2. *To cope with component and device tolerances in production, the circuit designer may provide a tuning slug in the coil.*

6-5 INTERSTAGE COUPLING WITH DOUBLE-TUNED NETWORKS

Advantages of double-tuned interstage coupling networks, as shown in Fig. 6-9, include a flatter frequency response within the pass

Interstage Coupling with Double-Tuned Networks · 147

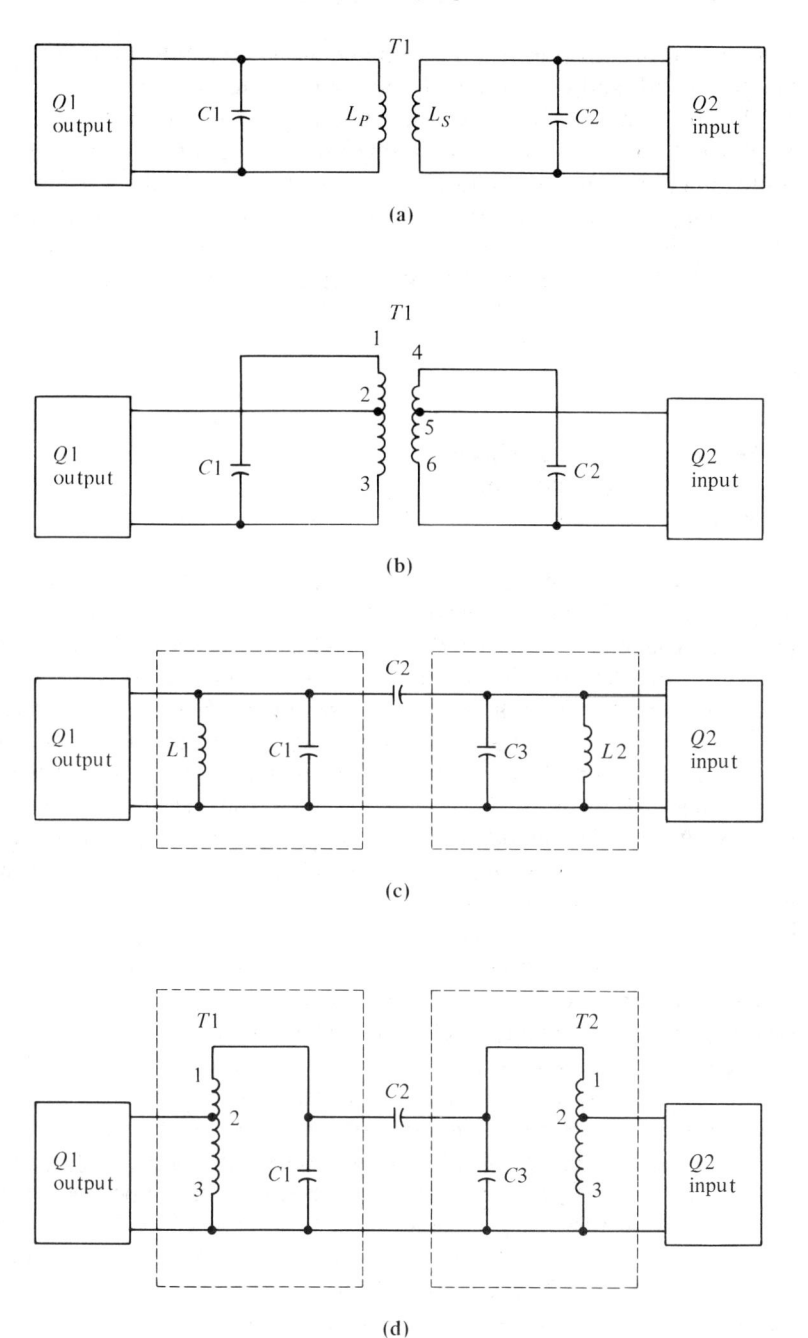

Figure 6–9 Double-tuned interstage coupling networks with inductive or capacitive coupling.

band, a sharper drop in response immediately adjacent to the ends of the pass band, and comparatively high attenuation of frequencies not in the pass band. Observe the frequency response curves exemplified in Fig. 6–10 for a double-tuned transformer with three coefficients of coupling. Inductively coupled circuits with a coefficient of coupling on the order of one percent are often used. When two coils are inductively coupled to provide transformer action as in Fig. 6–10, the coupling coefficient k is given by the equation

$$k = \frac{L_m}{\sqrt{L_1 L_2}}$$

where k = coupling coefficient.
L_m = mutual inductance.
L_1 and L_2 = inductance values of the coils.

Typical double-tuned transformers employ primary and secondary coils with a Q value of approximately 100. A k value of one percent provides a secondary bandwidth of 25 kHz. *If the circuit designer requires a somewhat greater bandwidth, a resistor of suitable value may be shunted across the primary and/or secondary.* Although resistance loading reduces the output signal amplitude, it has the advantage of avoiding excessive sag in the top of the response curve that results from using high values of k. Note in Fig. 6–10 that the maximum secondary current is obtained at critical coupling ($k = 0.002$). Next, observe in Fig. 6–11 that greater bandwidth at higher current output could be obtained by using coils with higher Q values. On the other hand, the sag in the top of the response curve would be greatly excessive. Again, if coils with lower Q values were used, the output current would be decreased. An optimum Q value results in a comparatively high output and a comparatively flat-topped curve.

Observe next the response-curve skirt contours exemplified in Fig. 6–12. A single-tuned coupling circuit provides comparatively poor selectivity and little approximation to a flat peak response. On the other hand, a double-tuned coupling transformer with a kQ product of 2 provides a reasonable approximation to flat-topped peak response, with a relatively rapid dropoff in response for frequencies past the maximum output point. Note that kQ products less than 2 result in progressively poorer selectivity and output-level characteristics. *The circuit designer should provide trimmer capacitors or tuning slugs in both the primary and the secondary windings to compensate for component and device tolerances.*

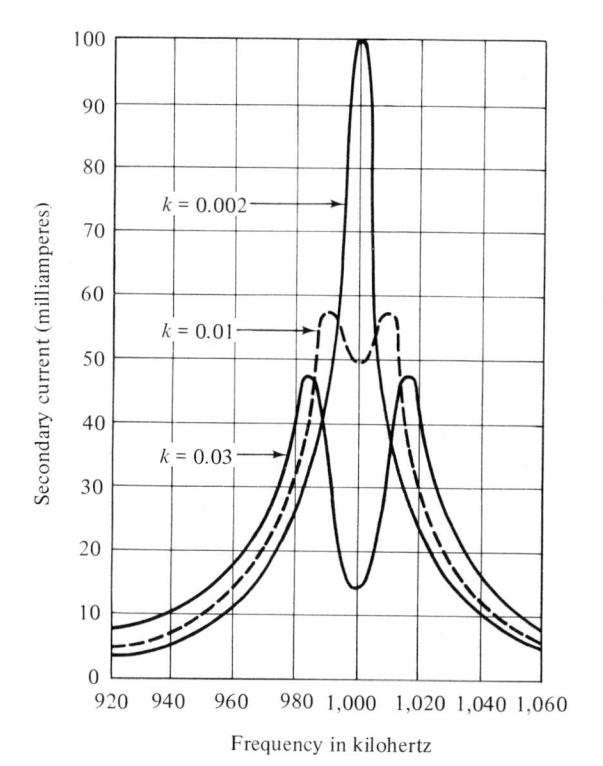

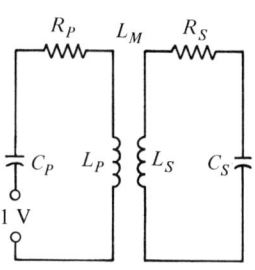

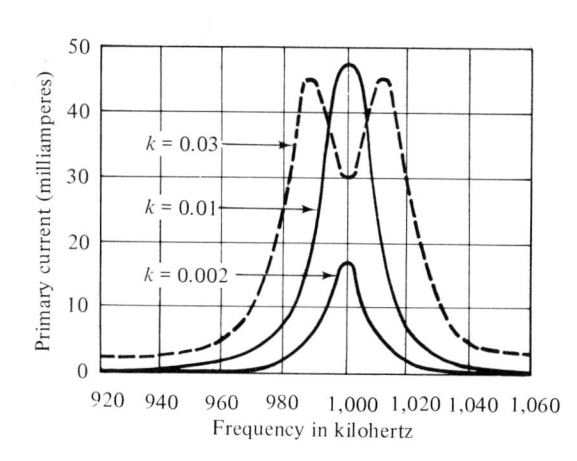

Interstage Coupling with Double-Tuned Networks · 149

Figure 6–10 Primary and secondary characteristics for a typical double-tuned transformer, for three coefficients of coupling.

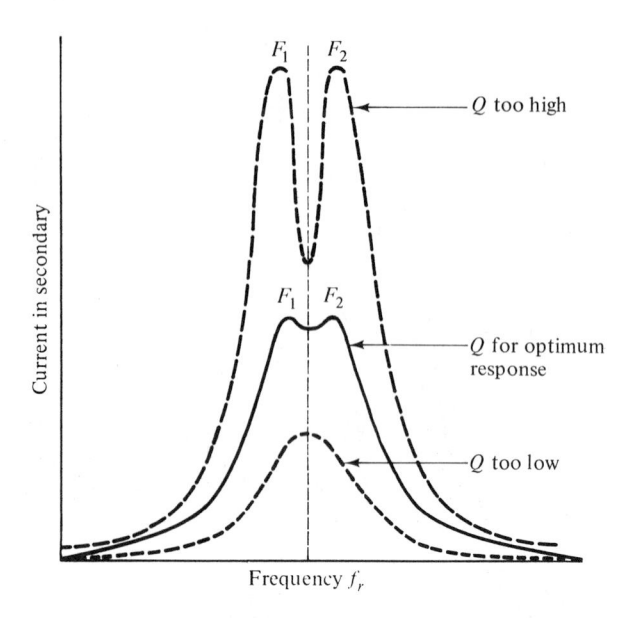

Figure 6–11 Response curves for a double-tuned transformer with three different Q values.

Consider the inductive-coupling arrangement shown in Fig. 6–9(a), and its alternative form depicted in Fig. 6–9(b). Capacitor C_1 and the primary winding L_p form a tuned circuit. Similarly, capacitor C_2 and the secondary winding L_s also form a tuned circuit. Each circuit is ordinarily tuned to the center frequency of the pass band; the double-humped responses in Fig. 6–10 result from overcoupling—not from stagger-tuning. However, designers sometimes choose a small amount of stagger-tuning to supplement a moderate amount of overcoupling. Impedance matching in Fig. 6–9(a) is obtained by proper selection of the primary and secondary turns ratio. When this configuration does not provide the required selectivity, the circuit designer may use tapped primary and secondary windings, as shown in Fig. 6–9(b).

Capacitive Coupling Observe the interstage coupling networks depicted in Fig. 6–9(c) and (d). Two capacitance-coupled tuned circuits are utilized. With reference to Fig. 6–9(c), capacitor C_1 and coil L_1 form a resonant circuit. Similarly, capacitor C_3 and coil L_2 form a resonant circuit. Each coil is tuned to the same frequency. Impedance matching is obtained by properly selecting the ratio of

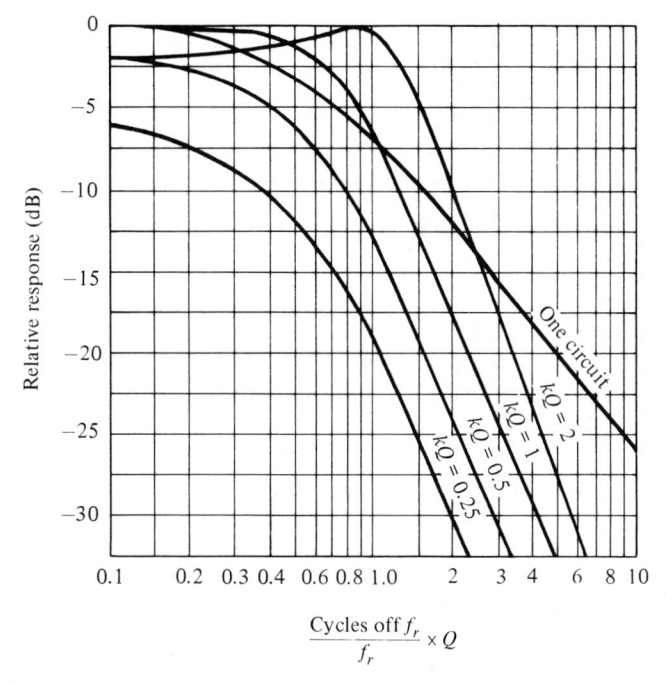

Figure 6–12 Response-curve skirt contours for single-tuned coupling circuit and for three different double-tuned coupling arrangements.

The graph's x-axis is labeled $\dfrac{\text{Cycles off } f_r}{f_r} \times Q$ and the y-axis is Relative response (dB). Curves shown include "One circuit", $kQ = 2$, $kQ = 1$, $kQ = 0.5$, and $kQ = 0.25$.

the reactance of capacitor C_2 to the impedance of the parallel input circuit (capacitor C_3 and coil L_2). The circuit shown in Fig. 6–9(d) functions in the same manner as noted previously; tapped coils (autotransformers) are utilized to facilitate the attainment of high selectivity. To compensate for production tolerances on components and devices, the circuit designer should provide each coil with a trimmer capacitor or with a tuning slug.

Gain Equalization Various applications of tuned amplifiers require that the center frequency of the amplifier be varied over a wide range of frequencies. Either capacitance variation or inductance variation may be employed. However, *the circuit designer is likely to encounter difficulty in maintaining constant gain over a wide frequency range* because the transistor beta tends to drop off more or less at higher operating frequencies. To compensate for this loss in gain at the higher frequencies, an equalizer circuit such as that shown in Fig. 6–13 may be utilized. This parallel combination

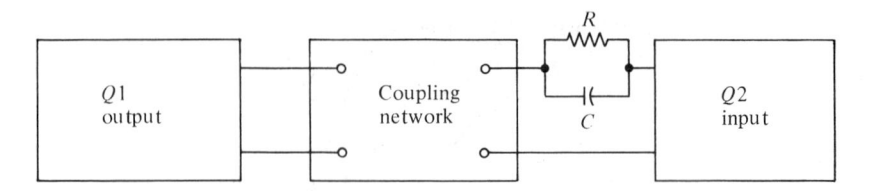

152 · Fundamentals of Radio-Frequency Circuit Design

```
┌─────────┐         ┌──────────┐    R
│         │────o  o─│          │───WWW───┐   ┌─────────┐
│   Q1    │         │ Coupling │──┤ (────┤   │   Q2    │
│ output  │         │ network  │    C        │ input   │
│         │────o  o─│          │─────────────│         │
└─────────┘         └──────────┘             └─────────┘
```

Figure 6–13   Coupled transistor amplifiers with an RC equalizing circuit.

of R and C is devised to attenuate the low frequencies more than it does the high frequencies. Thus, at high frequencies, capacitor C effectively bypasses resistor R, whereas at low frequencies the signal is attenuated by R.

6–6 NEUTRALIZATION AND UNILATERALIZATION

A *unilateral* electrical device can transmit energy in one direction only. A transistor is not inherently a unilateral device, because voltage variations in its output circuit cause voltage variations in its input circuit. In each transistor configuration, the feedback voltage from the output aids the input voltage, and is accordingly a *positive* feedback voltage. If the positive feedback voltage is sufficiently large, the amplifier will oscillate. At audio frequencies, the voltage feedback is low; therefore, precautions to prevent oscillation are not usually required. On the other hand, high-frequency amplifiers with tuned coupling circuits are susceptible to oscillation, and precautions must be taken by the circuit designer in various situations.

The effect of positive or negative feedback voltage on the input circuit of a device is to alter its input impedance. Usually, both the resistive and the reactive components of the input impedance are affected. The change in input impedance of a transistor caused by internal feedback can be eliminated by utilization of an *external* feedback circuit. If the external feedback circuit cancels both the resistive and the reactive changes in the input circuit, the transistor amplifier is said to be *unilateralized*. Alternatively, if the external feedback circuit cancels only the reactive changes in the input circuit, the transistor amplifier is said to be *neutralized*. Thus, neutralization is a special case of unilateralization. In either case, the external feedback circuit prevents oscillation in the amplifier system.

Unilateralized Common-base Amplifier   A tuned common-base amplifier configuration is shown in Fig. 6–14. DC biasing circuits

(a)

(b)

(c)

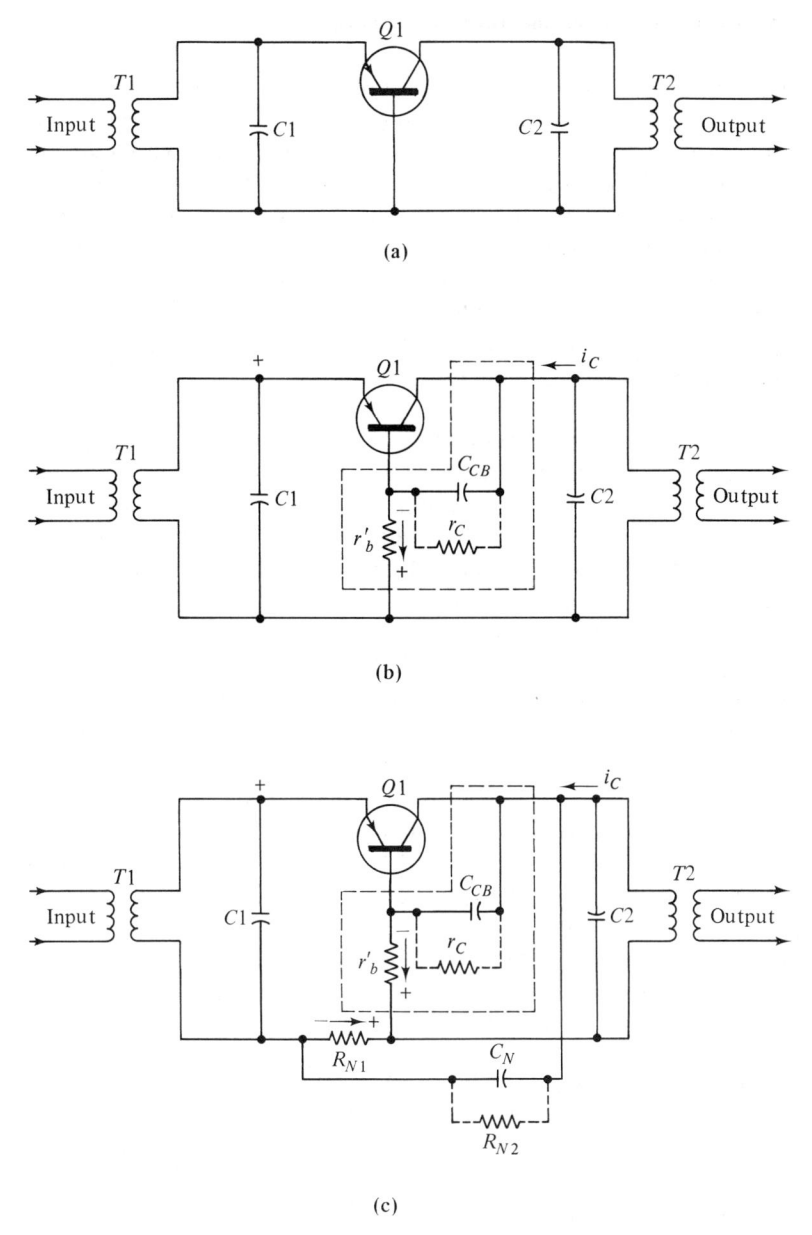

Figure 6-14 Common-base amplifier arrangements, showing internal feedback elements and external unilateralizing circuit.

are not shown. Transformer T1 couples the input signal to the amplifier. Capacitor C1 and transformer T1 secondary winding form a parallel-resonant circuit. Transformer T2 couples the output of the

153

amplifier to the following stage. Capacitor C2 and transformer T2 primary winding form a parallel-resonant circuit. The internal *elements* of the transistor that may cause sufficient positive feedback to result in oscillation are shown in dashed lines in Fig. 6–14(b). Resistor r'_b represents the resistance of the bulk material of the base, and is termed the *base-spreading resistance*. Capacitor C_{CB} represents the capacitance of the base-collector junction. Resistor r_c represents the resistance of the base-collector junction. This resistance is very high in value, because the collector junction is reverse-biased. At very high operating frequencies, capacitor C_{CB} effectively shunts resistor r_c.

Assume that the incoming signal aids the forward bias (causes the emitter to go more positive with respect to the base). Collector current i_c increases in the direction shown. A portion of the collector current passes through C_{CB} through resistor r'_b in the direction shown, and produces a voltage drop of the indicated polarity. The voltage across resistor r'_b aids the incoming signal, and therefore represents positive feedback that may cause oscillation. *Note in passing that if the circuit designer deliberately mismatches the tuned circuits to the transistor, oscillation can be prevented. However, this expedient results in a loss of signal power that may not be tolerable.*

When the incoming signal aids the forward bias, as noted above, the collector current increases. A portion of the collector current passes through C_{CB} through r'_b, in the direction shown, producing a voltage drop of the indicated polarity. A portion of the collector current also passes through C_N through R_{N1}, producing a voltage drop with the indicated polarity. The voltages across resistors r'_b and R_{N1} are opposing voltages. If the voltages are equal, no positive or negative feedback from the output circuit to the input circuit can occur. In turn, the amplifier is said to be *unilateralized*.

Common-emitter Amplifier, Partial Emitter Degeneration

A common-emitter amplifier configuration that utilizes partial emitter degeneration is shown in Fig. 6–15. Capacitor C_N and resistors R_{N1} and R_{N2} form a unilateralizing network. Transformer T1 couples the input signal to the base-emitter circuit. Resistor R1 forward-biases the base-emitter circuit. Capacitor C1 prevents short-circuiting of the base bias voltage by the secondary of T1. Transformer T2 couples the output signal to the following stage. Capacitor C2 and the transformer T2 primary winding form a parallel-resonant circuit. Capacitor C3 blocks battery DC voltage from R_{N1} and couples a portion of the collector current i_{c2} to the emitter. Inductor L1 is an RF choke that prevents bypassing of AC signal to ground via C3.

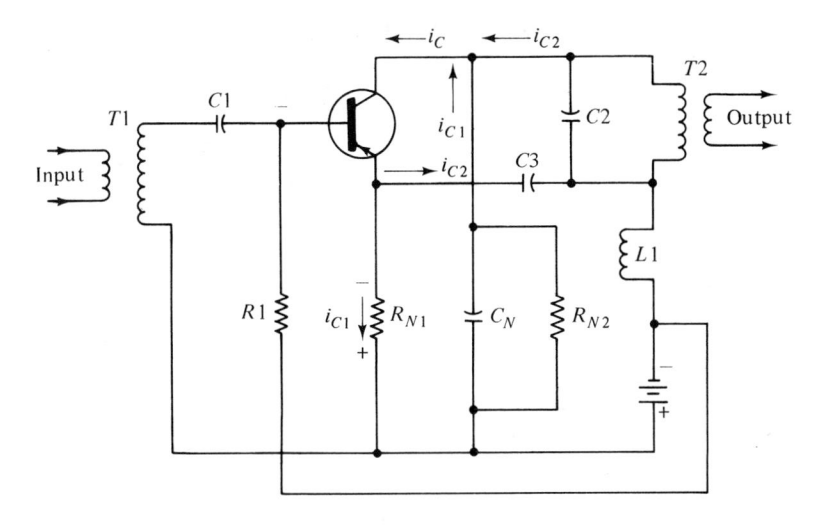

Figure 6-15 Common-emitter amplifier with partial emitter degeneration.

When the input signal aids the forward bias, collector current i_c increases in the direction shown. Internally, a portion of the collector current is coupled to the base-spreading resistance through the collector-base junction capacitance. The voltage developed across the base-spreading resistance aids the incoming signal and represents positive feedback. To offset this positive feedback, a portion of the collector current i_{c1} is directed through R_{N1} and the parallel combination of C_N and R_{N2}. The voltage developed across R_{N1} is a negative-feedback voltage equal and opposite that developed across the base-spreading resistance. Thus, the net voltage feedback to the input circuit is zero, and the amplifier is unilateralized. Note that the values of C_N, R_{N1}, and R_{N2} depend upon the internal values of collector-base junction capacitance, base-spreading resistance, and collector resistance, respectively.

It is essential to make a careful worst-case analysis of the design for any unilateralized or neutralized RF amplifier. Otherwise, a large number of rejects may occur in production owing to self-oscillation. In most cases, the circuit designer will make an experimental analysis by assembling breadboard versions of the amplifier with various combinations of selected components and devices that have positive-tolerance limits and others that have negative-tolerance limits.

Common-emitter Amplifier, Bridge Unilateralization A configuration for a common-emitter amplifier that is unilateralized by using transformer T2 winding 2-3 and the network consisting of R_N and C_N

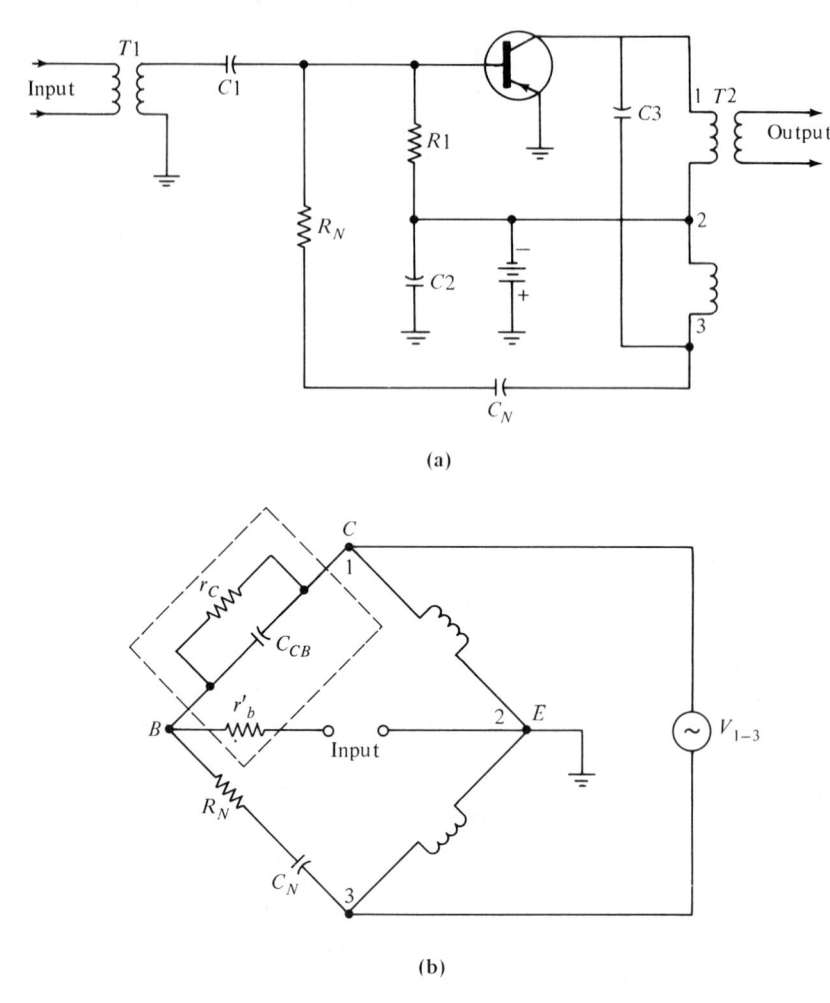

Figure 6–16 Common-emitter amplifier configuration, with bridge arrangement to prevent oscillation.

is shown in Fig. 6–16. Transformer T1 couples the input signal to the base-emitter circuit. Transformer T2 winding 1-2 couples the output signal to the following stage. Resistor R1 forward-biases the transistor. Capacitor C1 prevents short-circuiting of the base-bias voltage by the secondary of T1. Capacitor C2 bypasses the collector battery and places terminal 2 of T2 at AC ground potential. Capacitor C3 tunes the primary of T2. The operation of this circuit can be demonstrated by showing the bridge arrangement formed by the transistor internal feedback elements and the external unilateralizing network as seen in Fig. 6–16(b).

This bridge does not involve T1, C1, C2, or C3, R1, T2 secondary, or the collector battery. Points *B, C,* and *E* on the bridge represent the base, collector, and emitter terminals, respectively, of the transistor. Components shown in dashed lines are the transistor internal feedback elements. Points marked 1, 2, and 3 correspond to the terminals of the T2 primary. The voltage developed across terminals 1 and 3 of T2 is represented by a voltage generator with an output v_{1-3}. When the bridge is balanced, no part of voltage v_{1-3} appears between points *B* and *E.* In turn, the amplifier is said to be unilateralized. The bridge is balanced when the ratio of the voltages between points *B* and *C* and points *B* and 3 equals the ratio of the voltage between points *C* and *E* and points *E* and 3. In addition, the phase shift introduced by the network between points *B* and *C* must equal the phase shift introduced by the network between points *B* and 3.

In many applications, use of capacitor C_N is sufficient to prevent oscillation in the amplifier. In such a case, R_N is not used. Note that the amplifier is said to be neutralized (not unilateralized) in this case. Whenever one of these bridge methods is employed for control of positive feedback, *it is important for the circuit designer to make a careful worst-case analysis of the arrangement.* Tolerances on components and/or devices may have to be tightened to avoid the hazard of production rejects owing to self-oscillation. Since a comprehensive mathematical analysis of a bridge-unilateralized amplifier configuration is quite laborious, the circuit designer will usually make an experimental analysis by preparing breadboard versions of the arrangements using various combinations of components and devices with positive-limit and negative-limit tolerances.

6–7 CALCULATOR AIDED DESIGN

Impedance-matching networks comprising *L* or π sections with inductors and capacitors are utilized to obtain maximum power transfer, as from a radio transmitter to an antenna transmission line. Basic impedance-matching networks with required parameters are shown in Fig. 6–17. Note that the *Q* values in the matching networks must be selected in accordance with the equations that are employed. In Fig. 6–17(a), the *Q* value is equal to X_L/R at the operating frequency, or to R_{in}/X_C. In other words, after values of R_{in} and

158 · Fundamentals of Radio-Frequency Circuit Design

(a)

$R_{in} > R$

$X_L = \sqrt{RR_{in} - R^2}$

$X_C = \dfrac{R R_{in}}{X_L}$

(b)

$R_{in} < R$

$X_C = R\sqrt{\dfrac{R_{in}}{R - R_{in}}}$

$X_L = \dfrac{R R_{in}}{X_C}$

(c)

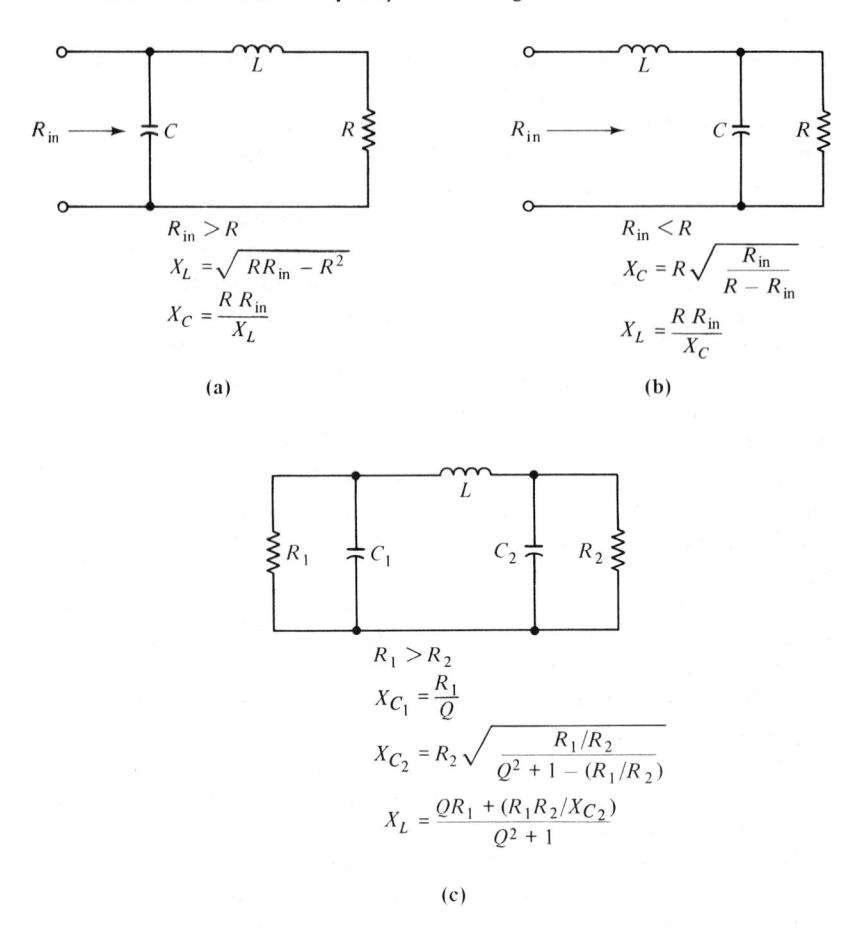

$R_1 > R_2$

$X_{C_1} = \dfrac{R_1}{Q}$

$X_{C_2} = R_2 \sqrt{\dfrac{R_1/R_2}{Q^2 + 1 - (R_1/R_2)}}$

$X_L = \dfrac{QR_1 + (R_1 R_2 / X_{C_2})}{Q^2 + 1}$

Figure 6-17 Basic impedance-matching networks: (a) L network, high Z to low Z; (b) L network, low Z to high Z; (c) π network, high Z to low Z.

R are assigned, specific values for L and C follow from the calculated values of X_L and X_C at the operating frequency. In Fig. 6-17(b), the Q value is equal to X_L/R_{in}, or to R/X_C. In Fig. 6-17(c), the Q value is equal to R_1/X_{C1}.

The π network depicted in Fig. 6-17(c) is in very extensive use. Although the associated equations can be solved with a slide rule, supplemented by pencil and paper, calculator-aided design can save appreciable labor for the circuit designer. Consider the keyboard for the calculator shown in Fig. 6-18. All trigonometric and logarithmic functions, square root, addition, subtraction, multiplication, division, and several other mathematical operations can be performed with a single keystroke. It has an "operational stack" of four

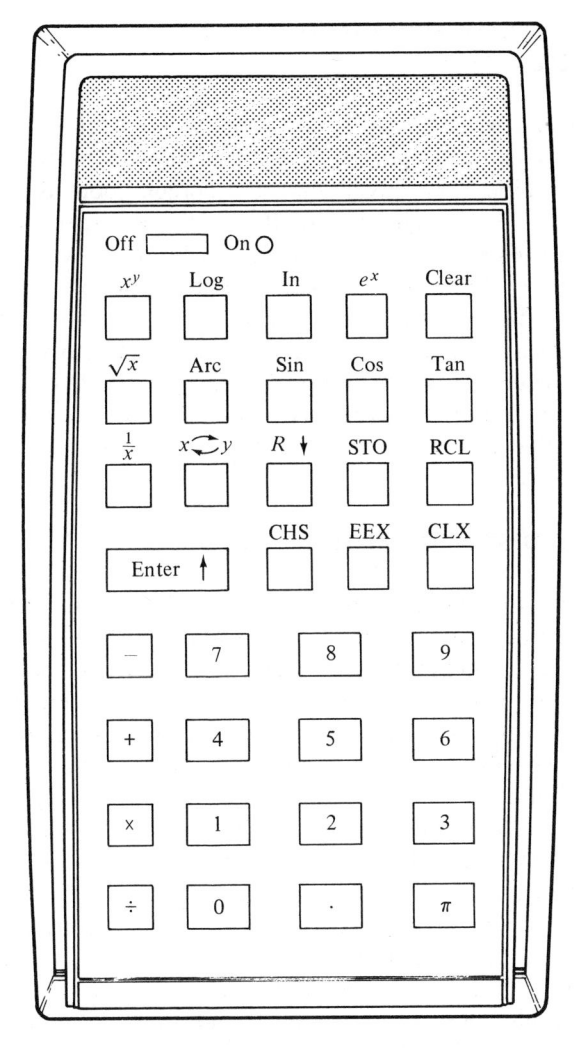

Figure 6-18 Keyboard of the Hewlett-Packard H-35 pocket calculator.

registers, plus a data storage register for constants. The stack holds intermediate answers and, at the appropriate time, automatically brings them back for further use. This eliminates the need for scratch notes or the reentry of intermediate answers when chains of calculations such as sums of products or products of sums are performed. In turn, the circuit designer can solve the pi matching network faster and with less labor on a calculator than with a slide rule and paper and pencil.

6-8 SCREEN ROOM ADVANTAGES IN RF DESIGN PROCEDURES

A screen room is a laboratory area that is completely enclosed by screen wire (usually copper), which is grounded in most instances to a cold-water pipe. The advantage of a screen room is its freedom from external electromagnetic radiation in the RF spectrum. In turn, low-level RF tests and measurements can be made without the hazard of false results owing to inadvertent RF interference. Power lines that run into a screen room are filtered at the point of entry by means of series RF chokes and shunt bypass capacitors grounded to the screen wire. No windows are ordinarily provided. The door is completely covered with screen wire, which in turn is bonded to the screen wire that covers the walls, floor, and ceiling.

6-9 GOOD DESIGN PRACTICES

Bypass capacitors utilized in solid-state RF circuits should be of mica or ceramic construction. Consideration must be given to low-impedance RF grounding arrangements. Copper-clad laminated circuit boards provide a good ground plane. Ground connections and connections to bypass capacitors or filter capacitors should be kept as short as possible. Chassis layout is planned to minimize ground paths and feedback coupling; stray capacitance and residual inductance are reduced insofar as may be practical. RF shields are used when necessary to avoid feedback coupling. Residual inductance can be reduced by employing very short leads; the emitter lead of a transistor is particularly important because of degeneration across its RF impedance. Hence, the emitter lead length must be minimized.

It is good practice to use a common ground return for each stage. A common RF ground point avoids the possibility of undesired coupling via ground loops. Adequate decoupling arrangements should be provided to prevent RF signal current from flowing through the power supply. Stray coupling between input and output leads can be minimized by separating them as much as possible, employing shields when necessary, and by dressing unshielded leads close to the chassis ground plane. Low-frequency parasitic oscillation can often be controlled by connecting a large bypass capacitor across the power-supply terminals. It is good practice to employ a plane chassis layout; in other words, if the ends of a chassis are bent around or folded, for example, the stray-coupling problem is likely to be intensified.

CHAPTER 7

Principles of Oscillator Circuit Design

7-1 GENERAL CONSIDERATIONS

All oscillator circuits employ regenerative (positive) feedback in an amplifier arrangement, as depicted in Fig. 7-1. Since the amplifier provides its own input, an output signal is developed that has a frequency equal to the resonant frequency of the tuned coupling circuit(s) in the amplifier. *Unless the oscillator circuit is elaborated, the output waveform is not a true sine wave.* In the first analysis, the amount of distortion in the output waveform is inversely proportional to the Q value of the tuned coupling circuit(s) in the amplifier. Note that the basic cause of distortion is the fact that positive feedback causes the oscillatory voltage in the amplifier to build up to a level at which device overload reduces the amplifier gain to nearly unity. This limiting overload action is accompanied by various forms of peak compression and/or clipping.

Some types of oscillators exploit clipping action to provide square-wave output. On the other hand, other types of oscillators, such as those utilized in high-fidelity audio oscillators, have design provisions to generate a low-distortion sine-wave output. Similarly, oscillators used in high-quality RF signal generators are designed to provide a low-distortion sine-wave output. This topic is concerned with the basic design principles of RF oscillators that employ LC resonant circuits. Frequency-determining elements in an oscillator circuit may consist of an inductance-capacitance arrangement, in single-circuit, double-circuit, or triple-circuit form, with either fixed

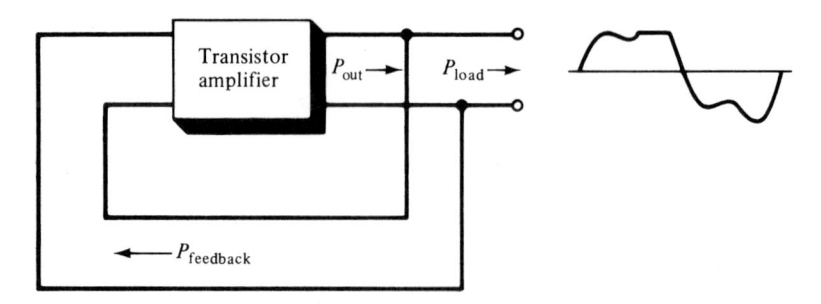

Figure 7-1 Block diagram of transistor oscillator, showing positive feedback of signal power output.

or variable tuning. Single-ended operation is generally utilized. *Push-pull operation reduces second-harmonic distortion,* and is used on occasion. Oscillator transistors operate in either the CE or the CB mode.

Bias voltage requirements for a transistor oscillator are similar to those for a conventional transistor amplifier. Observe that stabilization of the operating point is an important design factor in an oscillator circuit, because instability of the operating point will affect the output amplitude, waveform, and frequency stability. It is evident that frequency stability is a dominant consideration in design of local oscillators for radio and television receivers; a high order of frequency stability is also required in the design of oscillators for high-quality signal generators. *Circuit designers ordinarily employ regulated power supplies for operation of high-stability oscillators.*

7-2 TRANSISTOR IMPEDANCE CONSIDERATIONS

In a common-base configuration, the input impedance is low and the output impedance is high. To couple the feedback signal from the output to the input requires a feedback network that matches these unequal impedances. With practical limits, mismatch losses can be compensated by utilizing more feedback energy. In a common-emitter configuration, there is less difference between input and output impedance values, although impedance matching remains a design factor. The choice of configuration for an oscillator design is determined by the oscillator requirements and the advantages of a particular transistor amplifier configuration. As an illustration, *the alpha (common-base) cutoff frequency of a transistor is higher than is its beta (common-*

emitter) cutoff frequency. This margin in frequency-cutoff value is of assistance to the circuit designer in high-frequency oscillator operation.

Frequency Characteristics In general, the data sheet for a transistor includes cutoff frequency specifications. The minimum frequency capability of a transistor in a production lot is stated with relation to current gain. Of course, a production lot will contain transistors with a frequency capability appreciably greater than the rated value ("hot" transistors). Cutoff frequencies are designated as $f\alpha_b$ for the common-base configuration, and as $f\alpha_e$ for the common-emitter configuration. The cutoff frequency in either case is defined as the point at which the current gain drops 3 dB, or to 0.707 of its low-frequency value. Unless otherwise stated, this low-frequency value denotes 1-kHz operation.

Oscillators can operate efficiently above their cutoff frequency value, and can operate up to, but not including, the $f_{\max}$ rating of the transistor. This maximum frequency, or $f_{\max}$, rating is defined as the frequency at which the power gain of the transistor is unity. Since at least some power gain is required to overcome losses in the feedback circuit, oscillator operation at $f_{\max}$ is impossible. It is poor practice to design an oscillator for operation near the $f_{\max}$ rating of the transistor, because scant margin is then available for normal aging. In other words, *the circuit designer should anticipate a slow deterioration in the f_{max} value of a transistor after extended service.*

7–3 INITIAL AND SUSTAINED OSCILLATION

There is a basic distinction between self-excited and triggered oscillators. A self-excited oscillator starts to develop an output at the moment that DC power is applied. This output quickly builds up to a level at which the gain of the configuration is unity. On the other hand, a triggered oscillator does not develop an output until an input trigger pulse is applied. After the circuit has been triggered, the oscillatory output may decay to zero until another trigger pulse is applied. In some designs, the oscillatory output is self-sustaining after a trigger pulse has been applied. Closely related to these operating modes is *the hazard of oscillator dropout.* In other words, after the $f_{\max}$ value of a transistor has deteriorated past a critical limit, a variable-frequency oscillator will continue to develop output at the low-

164 · Principles of Oscillator Circuit Design

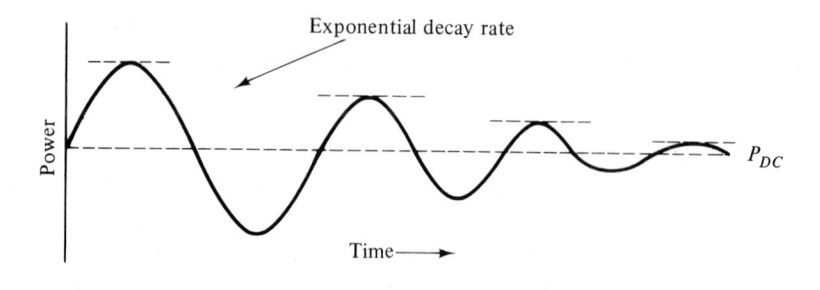

(a) Insufficient power feedback

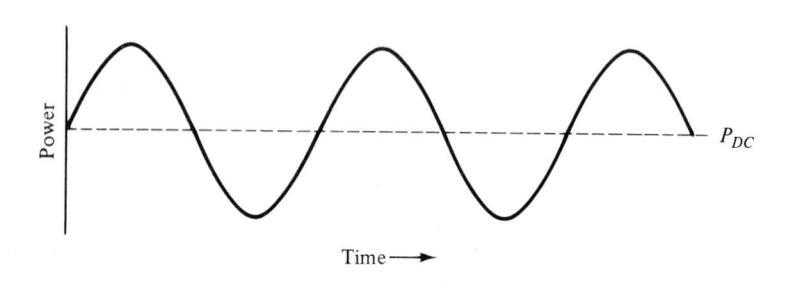

(b) Sufficient power feedback

Figure 7-2 Waveforms exhibiting the effect of sufficient and insufficient power feedback.

frequency end of its range, but "goes dead" or drops out at some frequency along the high end of its range.

The output power p_{out} of an oscillator is equal to the power gain of the amplifier section, multiplied by the input power (feedback power). If the amplifier gain is less than unity, and oscillation is initialized by any suitable means, a damped oscillatory output results, as depicted in Fig. 7-2. That is, a damped oscillation has progressively less amplitude and decays exponentially to zero. This output is characteristic of shock-excited oscillators, which may include an active device as a Q multiplier to provide regenerative action, or a simple resonant circuit may be utilized. If the power gain of the amplifier section is 0.9, and the initializing signal has a power of 1 dBm, the power output after the trigger pulse will be 0.9 dBm. If all of this power is fed back to the input and amplified, the power output on the second cycle becomes 0.81 dBm. Thus, the power output

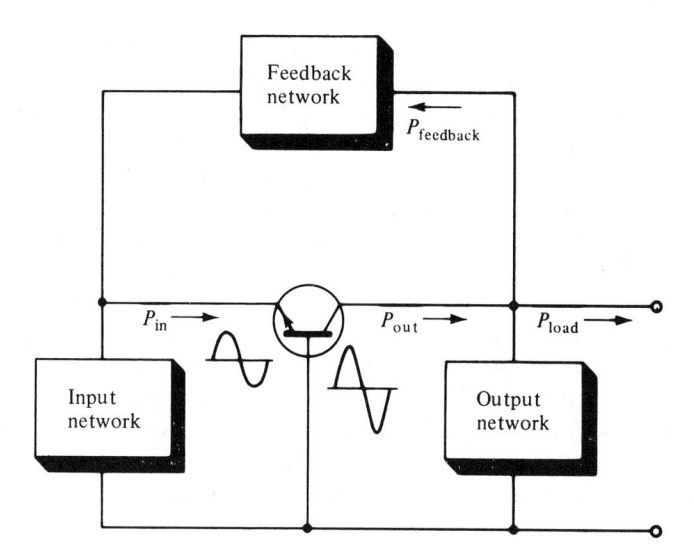

Figure 7-3 Block diagram of primary networks in a transistor oscillator arrangement.

progressively decreases, and approaches zero as a limit. Q multiplication results in a slower rate of decay than otherwise.

Next, consider the arrangement for self-sustained oscillation shown in Fig. 7-3. The power output is divided, as shown. One portion (p_{load}) is supplied to the load; the remainder ($p_{feedback}$) enters the feedback network. Power losses occur in the feedback network. Therefore, *the circuit designer should make a careful worst-case analysis of feedback network power losses.* For example, if the power gain of the transistor amplifier is equal to 20, and the input power p_{in} is 1 dBm, then the output power p_{out} is 20 dBm. If the amount of loss in the feedback network is 1 dBm, the feedback power $p_{feedback}$ must be 11 dBm to sustain oscillation. That is, $p_{feedback}$ less the loss in the network equals the p_{in} and is 1 dBm. Thus, all values remain constant, and oscillation is sustained. The power delivered to the load p_{load} is equal to the output power less the feedback power.

7-4 FREQUENCY STABILITY

It is instructive to consider the problem of frequency instability that is caused by transistor characteristics, rather than by external circuit elements. The DC operating point is usually chosen so that the operation of the transistor circuit is over the linear portion of its transfer characteristic. When circuit operation drifts to another operating point because of bias-voltage variation, the transistor parameters are shifted to a new set of values. Since these parameters are basic to the transistor, and affect the operating frequency, it follows that frequency drift will occur with bias-voltage variation. As noted previously, a constant supply voltage is a prime requirement for frequency stability.

Also, the collector-to-emitter capacitance parameter of the transistor affects the frequency stability more than the other parameters. This reactive element (sometimes called the barrier capacitance) varies with changes in collector or emitter voltages, and also with temperature. In high-frequency oscillators, when the collector-to-emitter capacitance becomes an important design factor, *the effect of capacitance changes may be minimized by inserting a relatively large swamping capacitor across the collector-to-emitter electrodes.* The total capacitance of the two in parallel results in an arrangement that is less responsive to voltage variations. Note that the added capacitor may be part of a tuned circuit.

It is evident that the use of a common bias source for both collector and emitter electrodes will maintain a relatively constant ratio of these two voltages. In effect, a change in one voltage is somewhat counteracted by the change in the other, since an increased collector voltage causes an increase in the oscillating frequency and an increased emitter voltage causes a decrease in the oscillating frequency. However, complete compensation is not obtained, because the effects on the circuit parameters of each bias voltage differ. In other words, it is a *conservative design practice* to operate a self-excited oscillator from a regulated power supply.

Temperature Considerations Changes in the transistor operating point with changes in temperature are a basic design consideration. Since the transistor oscillator is an amplifier with additional circuitry, the means of temperature stabilization that have been discussed for conventional amplifiers also apply to oscillator configurations. In addition, *the circuit designer will often find it necessary to include temperature-compensating capacitors in the frequency-determining circuit(s) of an oscillator.* In other words, fixed capacitors with various rated temperature coefficients are available, as shown in Fig. 7–4. In turn, any inherent tendency of the oscillator to drift in frequency with variation in temperature can be compensated (at least in part) by employment of one or more temperature-compensating capacitors in the tank

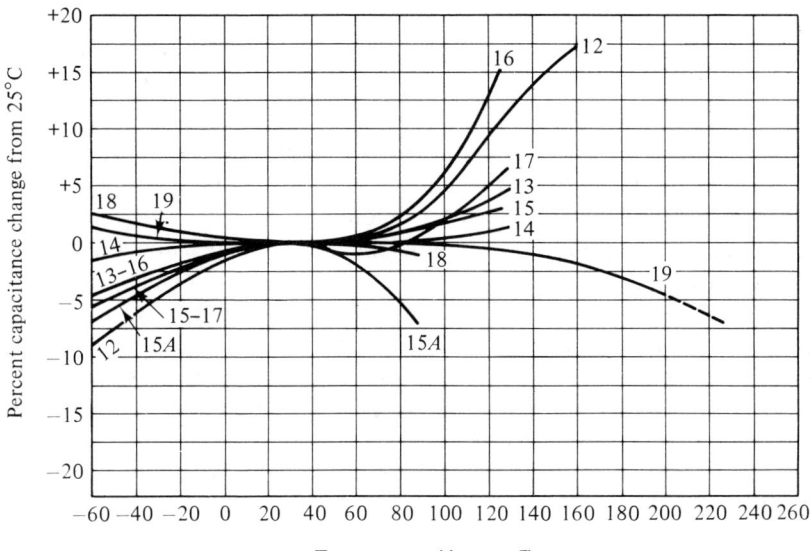

Frequency Stability · 167

[Graph: Percent capacitance change from 25°C vs Temperature (degrees C), showing curves labeled 12, 13, 14, 15, 15A, 16, 17, 18, 19]

12 Mylar
13 Paper Mylar
14 Polystyrene Mylar
15 Metalized Paper-Resin
15A Metalized Paper-Wax
16 Metalized Mylar
17 Metalized Paper Mylar
18 Polystyrene
19 Teflon

Figure 7-4 Capacitance-temperature characteristics for capacitors using various plastic films and combinations of dielectrics.

circuit(s). A temperature-compensating capacitor should be mounted near the oscillator transistor, so that their temperatures will "track."

It is seen in Fig. 7-4 that some temperature-compensating capacitors have a positive temperature coefficient; in other words, their capacitance value increases as the temperature increases. Other temperature-compensating capacitors have a negative temperature coefficient; in other words, their capacitance value decreases as the temperature increases. Still other temperature-compensating capacitors have a positive temperature coefficient at low temperatures and a negative temperature coefficient at high temperatures. This type of capacitor exhibits an increasing capacitance value from very low temperatures up to room temperature, and thereafter exhibits a decreasing capacitance value from room temperature to high temperatures.

If a capacitor has negligible capacitance change with temperature, it is termed an NPO type. On the other hand, if it has a negative temperature coefficient of capacitance, it may be identified as an N500 type, for example. This terminology means that the capacitor has a negative temperature coefficient and that its capacitance decreases 500 parts per million (ppm) per degree Celsius. Again, a capacitor that has a positive temperature coefficient may be identified as a P350 type, for example. This terminology means that its capacitance increases 350 ppm/°C. Note in passing that most RF coils have a positive temperature coefficient of inductance. In turn, this characteristic is taken into account when temperature-compensating capacitors are selected. This selection is ordinarily made on an experimental basis.

7–5 BASIC TRANSISTOR OSCILLATOR CONFIGURATIONS

Skeleton component and device arrangements with polarity indications for various transistor oscillators are shown in Fig. 7-5. Each of these circuits provides amplification, regenerative feedback, and *LC* tuning. Exact bias and stabilization arrangements are not shown, although the relative transistor DC potentials required for normal oscillator function are indicated. The circuits depicted in Fig. 7–5(a) and (b) are similar. In both circuits, feedback energy is coupled from the collector to the base by a transformer (often called a tickler-coil arrangement). The circuit in (b) is a shunt-fed version of the series-fed circuit in (a). Since the feedback signal proceeds from collector to base in each case, *the necessary feedback signal phase is obtained by specifying transformer connections that provide 180-deg phase shift.* If the connections to either coil are reversed, the configuration will not oscillate.

"Tickler" feedback circuits are also depicted in Fig. 7–5(c) and (d). In the (c) arrangement, regenerative feedback with zero phase shift is obtained in the tuned collector-to-emitter circuit by appropriate connections of the transformer terminals. In the example of (d), since the feedback is from collector to base, 180-deg phase shift is required. That is, the signal voltage in the untuned collector winding is coupled and inverted in phase into the tuned-base winding. In the first analysis, the oscillating frequency is determined by the reso-

Basic Transistor Oscillator Configurations · 169

Figure 7-5 Basic transistor oscillator configurations.

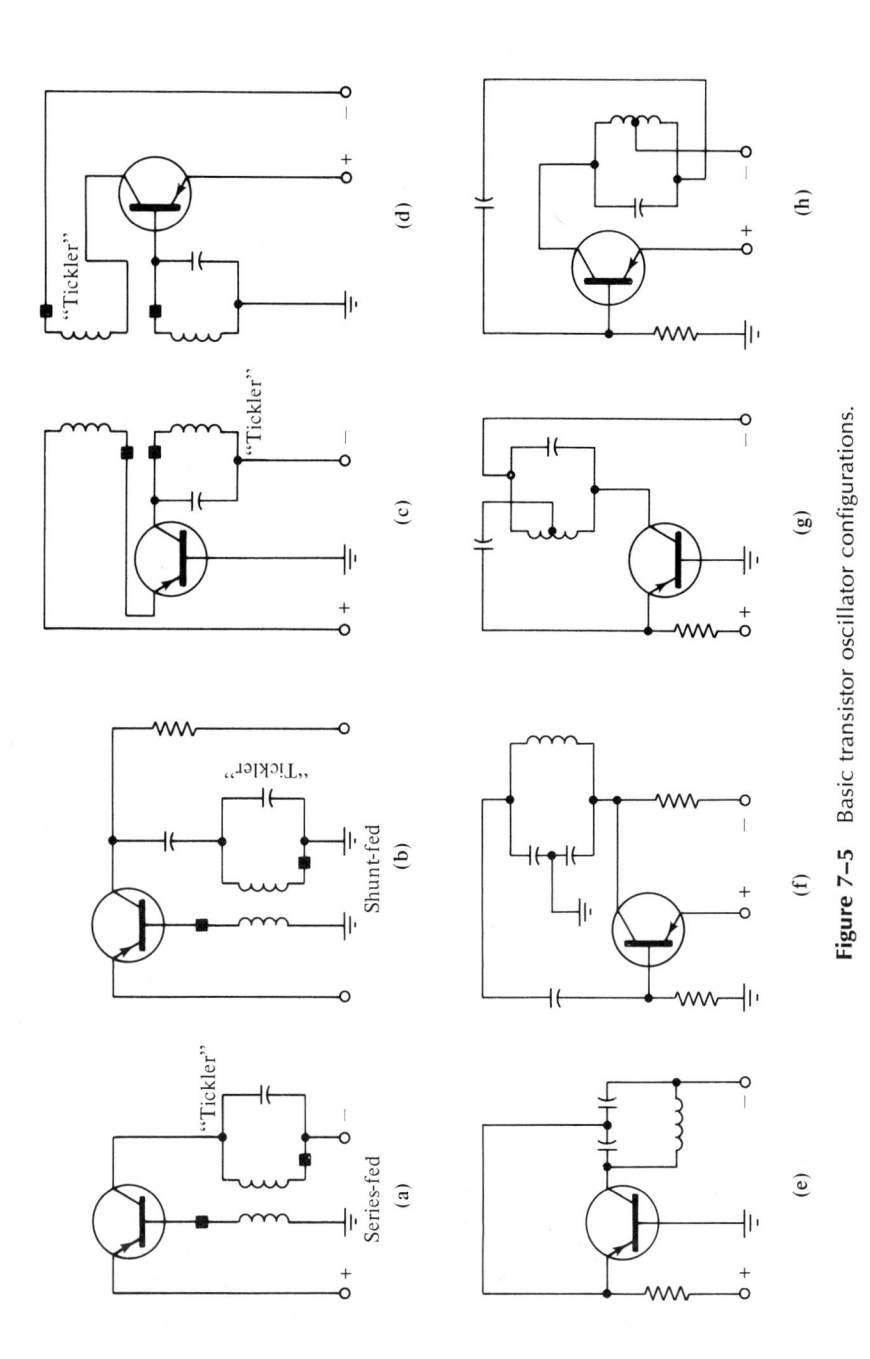

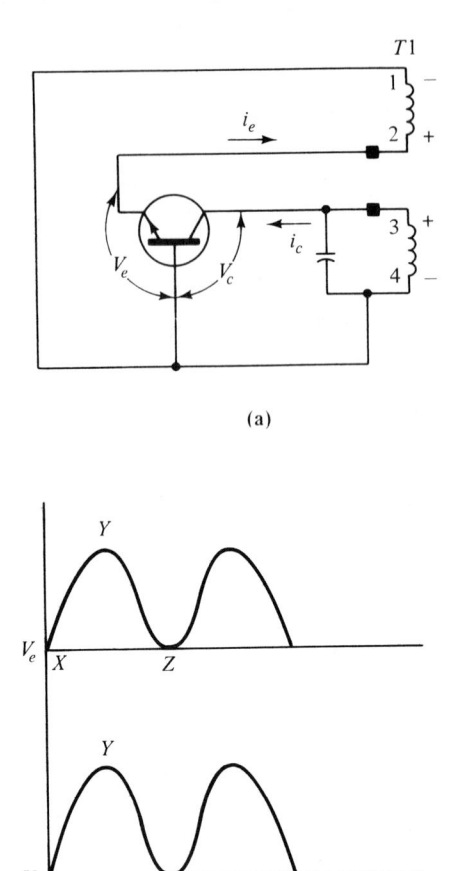

Figure 7-6 Skeleton transistor oscillator circuit, with voltage waveforms.

nant frequency of the parallel *LC* circuit. Although the untuned winding reflects an impedance into the tuned winding, this is a second-order factor. Therefore, *the circuit designer generally inserts temperature-compensating capacitors into the parallel-resonant circuit.* A temperature-compensating capacitor may be shunted across the tuning capacitor, or it may be connected in series with the coil and the tuning capacitor. A combination of both arrangements is sometimes required.

The oscillator circuits shown in Fig. 7–5(e) and (f) are transistorized versions of the standard Colpitts configuration. In (e), the signal from the tuned collector is coupled in phase into the emitter circuit. On the other hand, in (f), the signal from the tuned collector is coupled 180 deg out of phase into the base circuit. The circuits shown in Fig. 7–5(g) and (h) are similar to those depicted in (e) and (f), except that a split (tapped) inductor is utilized to provide the necessary feedback. These are transistorized versions of the standard Hartley configuration. In each of these circuits, the collector is tuned. Since each coil functions as an autotransformer, feedback in the correct phase is accomplished by inductive coupling. In the (g) configuration, the feedback signal is coupled from collector to emitter with no phase shift. On the other hand, in the (h) arrangement, the feedback signal is coupled from collector to base with a 180-deg phase shift.

Tickler-coil Oscillator Operation Consider the transistor oscillator circuit with a tickler coil for inductive feedback, depicted in Fig. 7–6(a). The nonlinear conditions for amplitude limiting in this circuit are based on a drive amplitude that causes the base-emitter junction to become reverse-biased (cut off), and a drive amplitude that causes the collector-base junction to become forward-biased (saturated). As detailed subsequently, overdrive past cutoff results in base-emitter rectification of the drive signal, and, in some oscillator configurations, results in self-bias (signal-developed bias) voltage that provides class C operation of the transistor. Note that an oscillator cannot operate in class A, unless some nonlinear negative-feedback action is included to control the buildup of the oscillator signal amplitude in the system. However, a simple oscillator arrangement may operate in class B.

Because class B and class C operation of a transistor is associated with nonsinusoidal current waveforms, it might be supposed that the output voltage waveforms are also nonsinusoidal. *However, this is not necessarily true.* For example, if the tank coil has a very high Q value, and in the absence of substantial output loading, the collector voltage waveform will be approximately sinusoidal. Again, if the base-emitter coil has a very high Q value and if it is self-resonant at the operating frequency, the emitter voltage waveform will also be approximately sinusoidal. In other words, high-Q coils have a *flywheel effect* owing to their energy storage, and they tend to maintain a sinusoidal terminal voltage although subjected to pulsed current flow.

172 · Principles of Oscillator Circuit Design

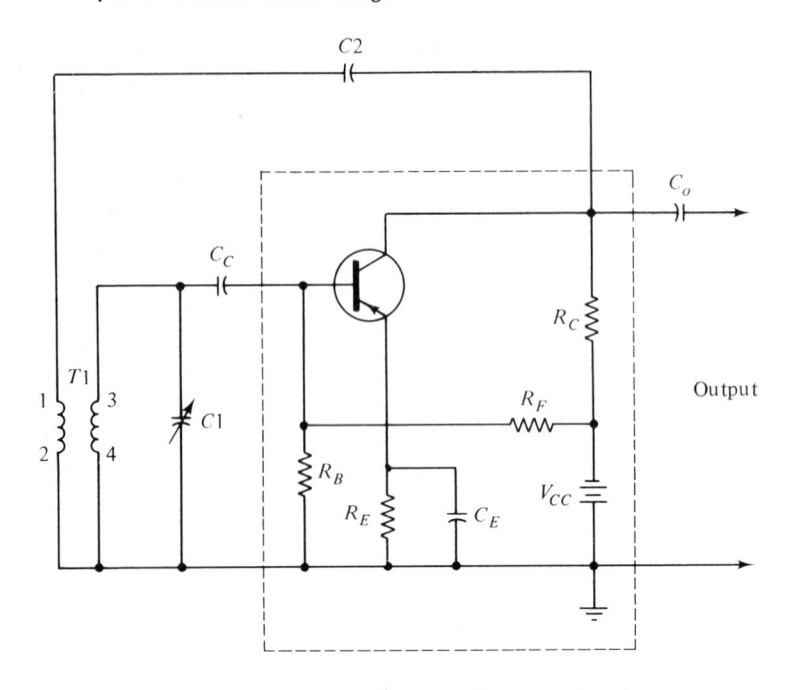

Figure 7-7 Basic tuned-based oscillator configuration.

With reference to Fig. 7-7, the tuned-base oscillator configuration exemplified has a shunt-feed collector circuit. Shunt feed obviates the need for DC current flow through tickler winding 1-2 of transformer T1. Positive feedback occurs as the result of mutual inductance between the transformer windings. Bias circuitry is the same as explained previously for amplifier design and operation. Note that R_E is part of the bias network (emitter swamping resistor) and that it is bypassed by C_E in order to realize maximum power gain. Since a base coupling capacitor (C_C) is included in the configuration, *signal-developed bias can become a significant factor in oscillator operation,* particularly when the windings on T1 are tightly coupled. In other words, base-emitter rectification of the drive signal can build up a substantial charge on C_C to reverse-bias the base-emitter junction. In this situation, the oscillator operates in class C; a DC voltmeter measurement of the base-emitter bias voltage will *seem* to indicate that the transistor is completely cut off. Of course, the base-emitter junction conducts in pulses at the peaks of drive.

Signal-developed bias can serve a particularly useful function in

Basic Transistor Oscillator Configurations · 173

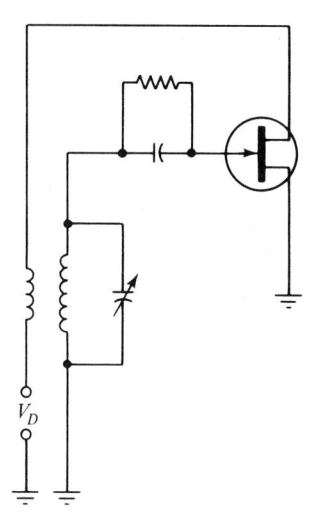

Figure 7-8 A JFET variable-frequency oscillator configuration that employs signal-developed bias.

the design of variable-frequency oscillators (VFO's). A junction field-effect transistor (JFET) variable-frequency oscillator configuration that utilizes signal-developed bias is exemplified in Fig. 7-8. A gate coupling capacitor and resistor are connected in series with the gate lead to develop signal-developed bias. Overdrive of the transistor by the feedback voltage results in gate-source rectification of the drive signal that builds up a charge on the coupling capacitor and reverse-biases the gate. The magnitude of this reverse bias depends upon the amplitude of the oscillator signal in the drain circuit. In turn, *as the tuning capacitor is varied through its range, any tendency of the output signal amplitude to decrease is counteracted by reduced gate-bias voltage, and increased transistor gain. Thus signal-developed bias assists the circuit designer in maintaining a constant output amplitude from a VFO.*

Consider next the basic tuned-collector oscillator configuration depicted in Fig. 7-9. Resistors R_F and R_B establish the base bias voltage. Resistor R_E is an emitter swamping resistor. Capacitors C_B and C_E bypass the RF signal around R_B and R_E, respectively. Series feed is utilized. The tuned circuit comprises transformer winding 3-4 and variable capacitor C1, which provides VFO operation. Oscillation starts as soon as DC power is applied. Signal-developed bias occurs in consequence of base-emitter signal rectification and charge

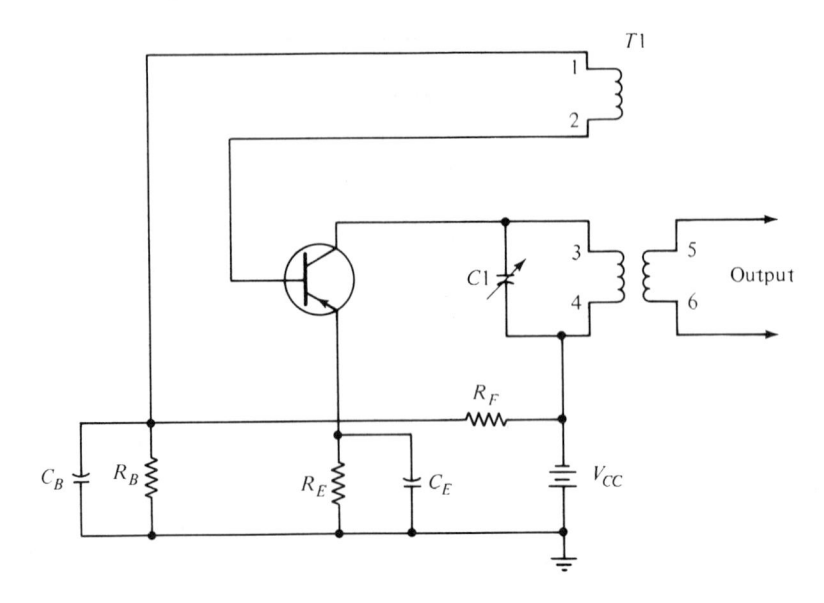

Figure 7-9 Basic tuned-collector oscillator configuration.

buildup on C_B and C_E. Positive feedback takes place from transformer winding 3-4 to winding 1-2. Note that transformer winding 5-6 couples the oscillator output signal to a load. *As a general rule, circuit designers use link coupling for output to a low-impedance load, but utilize capacitive coupling for output to a high-impedance load.*

7-6 COLPITTS AND CLAPP OSCILLATOR OPERATION

A transistor Colpitts oscillator configuration is illustrated in Fig. 7-10. Regenerative feedback is taken from the tank circuit and is applied to the emitter of the transistor. Base bias is provided by resistors R_B and R_F. Resistor R_C is the collector load resistor. Resistor R_E develops the emitter input signal and also functions as the emitter swamping resistor. The tuned circuit comprises capacitors C1 and C2 in parallel with transformer winding 1-2. Note that capacitors C1 and C2 form a voltage divider. The voltage developed across C2 is the positive feedback voltage. Either or both capacitors may be adjusted to control the frequency and the amount of feedback voltage. *For minimum feedback loss, the ratio of the capacitive*

Colpitts and Clapp Oscillator Operation · 175

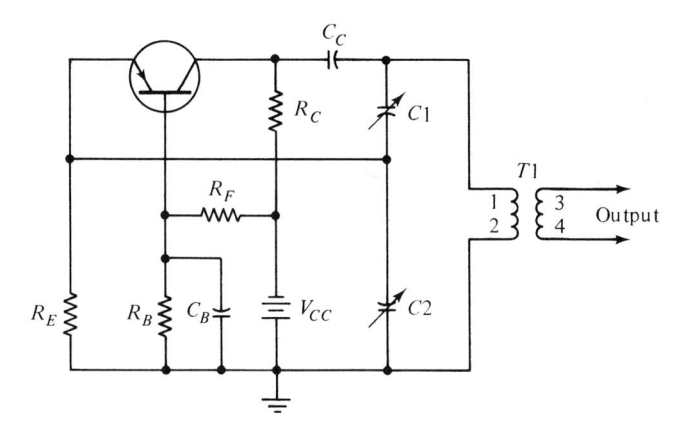

Figure 7-10 Typical Colpitts oscillator configuration.

reactances of C1 and C2 should be approximately equal to the ratio of output impedance to the input impedance of the transistor. Since shunt feed is employed, the tank circuit is at zero DC potential. Link-coupled output to a low-impedance load is indicated in this example.

Clapp Oscillator Operation Modification of the Colpitts oscillator arrangement by inclusion of a capacitor in series with winding 1-2 of the transformer results in a Clapp oscillator configuration, as exemplified in Fig. 7-11. *This added capacitor improves the frequency stability of the oscillator system.* When the series capacitance of C1 and C2 is large compared with the added capacitance, the oscillating frequency is effectively determined by transformer winding 1-2 with C in series. Capacitor C is made variable for VFO operation. At the resonant frequency, the shunting impedance of the LC series combination has a minimum value, making the oscillating frequency comparatively independent of transistor parameters.

Simplified Feedback Networks The common-base configuration is used in the Colpitts oscillator exemplified in Fig. 7-10. No phase shift exists between the signal on the collector and the signal on the emitter. Therefore, the feedback signal is not shifted in phase. The common-emitter configuration is utilized in the Clapp oscillator exemplified in Fig. 7-11. A 180-deg phase shift is provided between the signal on the collector and the signal on the base. In this instance, the feedback signal is inverted to provide an in-phase rela-

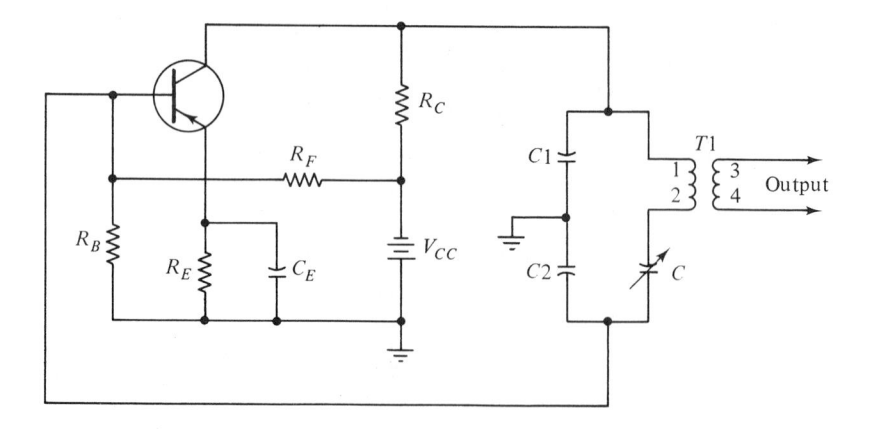

Figure 7-11 Typical Clapp oscillator configuration.

tionship at the input. When the CB configuration is employed, the feedback signal is returned between emitter and ground, as depicted in the skeletonized circuit of Fig. 7-12(a). Only AC relations of the feedback signal are illustrated in the diagram. That is, the polarities indicated at the transistor and capacitor are instantaneous potentials. As the emitter goes negative, the collector also goes negative, developing the potential polarities indicated across C1 and C2. Feedback voltage is developed across C2, which is fed back between emitter and ground; it also goes negative. Thus, the in-phase relationship at the emitter is maintained.

If the circuit designer requires a very high order of frequency stability, the frequency-determining network should be isolated from the amplifier section by means of a buffer stage. This approach cushions the frequency-determining circuit from the inevitable changes induced by electrical variations, but requires higher losses in the coupling arrangement and necessitates a higher amplifier gain to satisfy stability criteria. An alternative to this elaboration is to employ a very high Q value in the resonant circuit. Observe next that when the CE configuration is utilized, as shown in the skeletonized diagram of Fig. 7-12(b), the feedback signal is returned between base and ground. As the base goes positive, the collector goes negative, developing the potential polarities across C1 and C2 as shown. The voltage across C2 is fed back between base and ground. Thus, the in-phase relationship at the base is maintained.

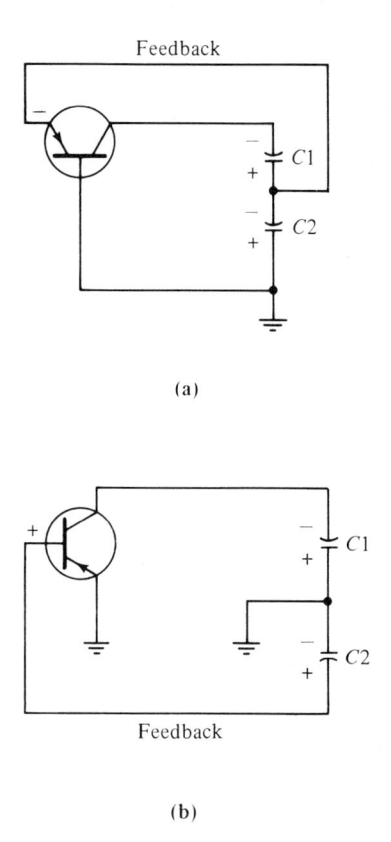

Figure 7-12 Skeletonized feedback networks for the Colpitts and Clapp oscillator configurations: **(a)** Colpitts circuits; **(b)** Clapp circuit.

7-7 HARTLEY OSCILLATOR OPERATION

Two versions of the basic Hartley oscillator are shown in Fig. 7-13. The shunt-fed and the series-fed versions are operationally similar. *An operating efficiency of over 50 percent can be obtained by proper design of the series-fed configuration.* A similar operating efficiency can be obtained with a slightly elaborated design of the shunt-fed configuration. With reference to the shunt-fed arrangement, R_B, R_C, and R_F provide suitable bias for the circuit. *Bias should be sufficient to ensure easy starting, and may require some sacrifice of operating efficiency.* The frequency-determining network consists of the series combination of transformer windings 1-2 and

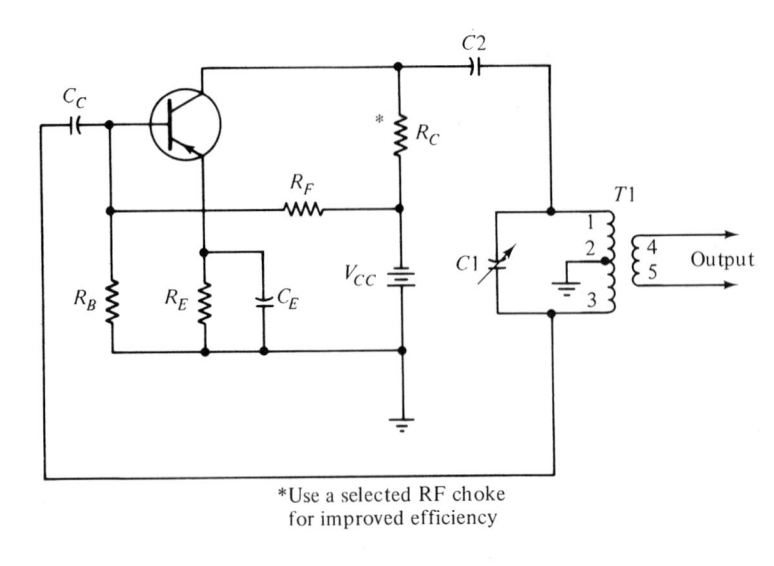

178 · Principles of Oscillator Circuit Design

*Use a selected RF choke for improved efficiency

(a)

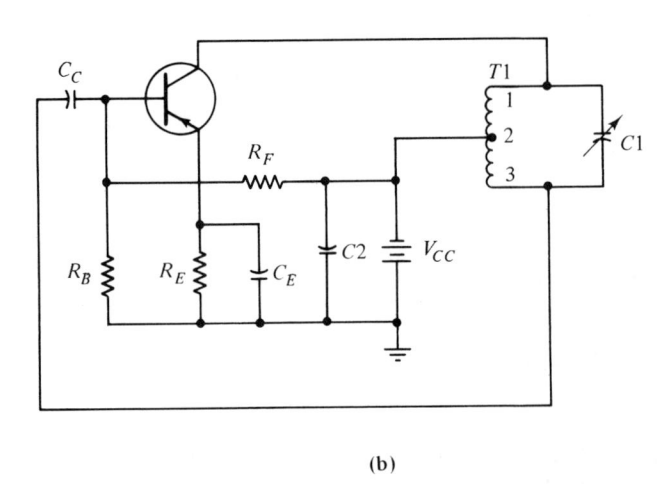

(b)

Figure 7-13 Basic Hartley oscillator arrangements: **(a)** shunt-fed; **(b)** series-fed.

2-3 in parallel with C1. VFO operation is provided by making C1 variable. Capacitor C2 is a DC blocking capacitor. Capacitor C_E provides a signal bypass around the emitter swamping resistor R_E. If the oscillator is designed to provide appreciable power output, R_C

should be replaced by a selected RF choke to maximize the operating efficiency.

In the circuit of Fig. 7–13(a), the inductance functions as an autotransformer to provide the regenerative feedback signal. This feedback is obtained from the induced voltage in transformer winding 2–3, coupled via C_C to the base of the transistor. Shunt feed avoids the necessity for passing DC current through the T1 primary winding. Link coupling to a low-impedance load is provided by winding 4–5. Next, in the series version of Fig. 7–13(b), R_B and R_F provide the necessary bias for the base-emitter circuit. Collector voltage is obtained through transformer winding 1–2. Capacitor C_E provides a signal bypass path around the emitter swamping resistor R_E and increases the power gain of the amplifier section.

Regenerative feedback is obtained from the induced voltage in transformer winding 2–3 and is coupled via C_C to the base of the transistor. Capacitor C2 places terminal 2 of transformer T1 at AC ground potential. If a low-impedance load is to be driven, a link coil comprising a few turns tightly coupled to winding 1–2 may be used. Or, if a high-impedance load is to be driven, the designer will usually provide capacitive coupling from terminal 1 of the transformer to the load. As a practical note, almost any *LC* oscillator circuit that is operated at a high frequency can be effectively analyzed in terms of either a basic Colpitts or Hartley configuration.

Push-pull Oscillator Operation A modified version of the basic Hartley oscillator configuration that operates in push-pull is exemplified in Fig. 7–14. As noted previously, an advantage of push-pull operation is elimination of second-harmonic distortion. However, *a trade-off is involved, in that third-harmonic distortion is increased in the push-pull operating mode.* In some design situations, push-pull operation of an oscillator has the advantage of greater power output than can be provided by a single transistor. Base bias for this circuit is established by R_B and R_F. Capacitor C_E provides a signal bypass path around the emitter swamping resistor R_E. The oscillating frequency is essentially determined by the tank circuit comprising transformer winding 4–6 and capacitor C2. Positive feedback is applied between base and emitter of each transistor by the induced voltages in transformer windings 1–3 and 4–6. Capacitor C1 places terminal 2 of T1 at AC ground potential via C_E.

It is comparatively difficult to design RF inductors that have a low temperature coefficient of resistance. Accordingly, the positive temperature coefficient of the inductor is ordinarily compensated

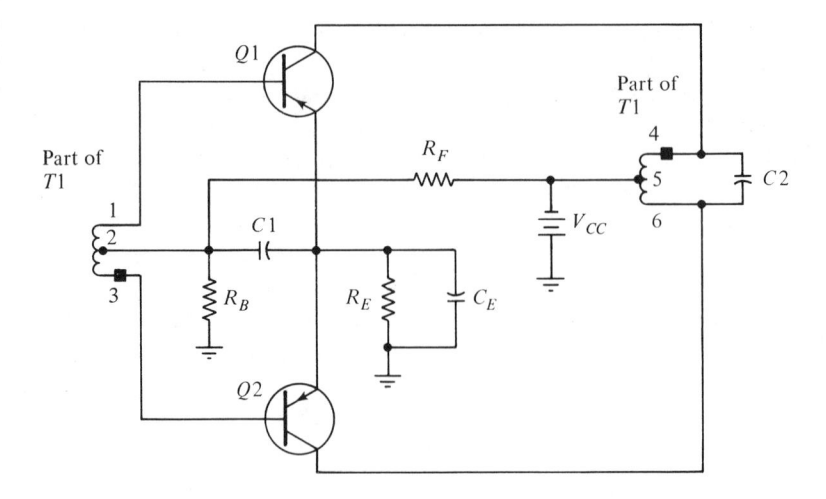

Figure 7-14 Example of a push-pull oscillator arrangement.

with capacitors that have a negative temperature coefficient. This method is usually satisfactory over comparatively limited temperature ranges. On the other hand, the design problem can become formidable over wide temperature ranges. Although the positive and negative temperature coefficients can be made to track in a prototype model, *it may be found impracticable to make the prototype design reproducible in large-scale production.* In this situation, the circuit designer will choose a more costly type of oscillator arrangement, such as a quartz-crystal configuration.

7-8 CRYSTAL OSCILLATORS

A quartz crystal has an advantage of very high Q values in oscillator operation. Q values range from 20,000 to 1,000,000. A suitably cut crystal has good frequency stability over a specified temperature range. For example, a GT crystal has a temperature coefficient of practically zero from zero deg to 100°C. A typical X-cut crystal has a temperature coefficient of -25 to $+100$ Hz/°C/MHz. X-cut crystals are suitable for operation at higher frequencies than are GT-cut crystals. When the tolerance on operating frequency is extremely tight, the crystal is maintained at constant temperature in a crystal oven. A quartz-crystal symbol and its equivalent circuit with a typi-

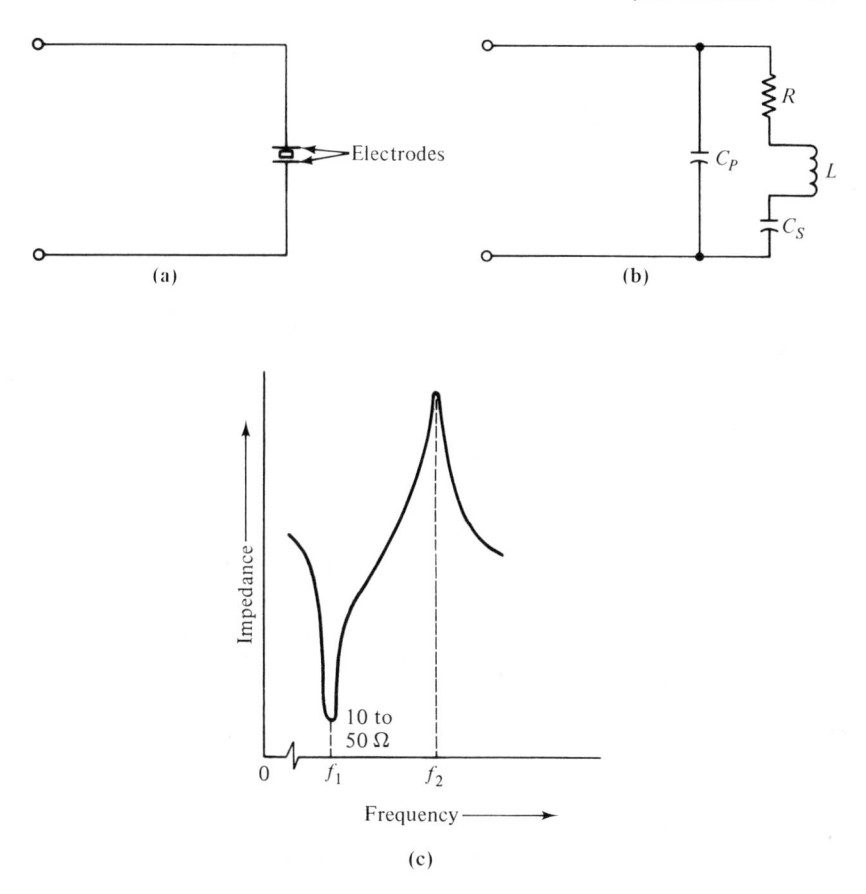

Figure 7–15 Quartz crystal characteristics: (a) arrangement in holder; (b) equivalent circuit; (c) impedance characteristic.

cal impedance characteristic are shown in Fig. 7–15. Resistance R, inductance L, and capacitance C_S in series, depicted in (b), represent the electrical equivalent of the mechanical vibrating characteristic of the crystal.

At series resonance of C_S and L, the impedance of the crystal is low, and the resonant frequency of the oscillator circuit is determined only by the mechanical vibrating characteristics of the crystal. The equivalent series resistance of the crystal is very low, and ranges up to 30 ohms in low-cost high-R crystals. Parallel capacitance C_P represents the capacitance between the crystal electrodes in Fig. 7–15(a). Above the frequency of series resonance, X_L is

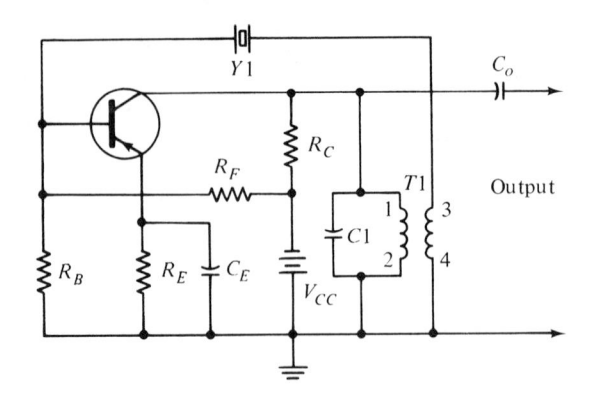

Figure 7-16 Crystal-controlled "tickler" feedback oscillator.

greater than X_{Cs}. The resulting net inductance forms a parallel-resonant circuit with C_P and any capacitance in shunt to the crystal. Thus, *the oscillating frequency can be changed within a limited range by means of a trimmer capacitor connected across the crystal holder.* Alternatively, a suitable inductor can be connected in series with the crystal holder. In parallel-resonant operation, the impedance of the crystal is high.

A quartz crystal may be operated in either its series-resonant or its parallel-resonant mode. Refer to Fig. 7-15(c). The slopes of both the series-resonant and the parallel-resonant curves are very steep, and indicate that the crystal has a very high Q value in either mode. For most crystals, the difference in frequency between f_1 and f_2 is very small, compared with the series-resonant frequency of the crystal. Note that the mode of operation is primarily determined by the impedance of the circuit to which the crystal is connected. *Whenever an oscillator has more than one possible frequency of operation, oscillation will always take place at the frequency corresponding to the highest Q value of the system.*

A tickler-coil crystal-controlled oscillator configuration is shown in Fig. 7-16. This circuit employs the series mode of operation for the crystal and is basically a tuned-collector type of oscillator. The oscillating frequency is determined primarily by the crystal, although the circuit capacitance in shunt to the crystal and the circuit inductance in series with the crystal have a small effect on the oscillating frequency. A small variation in frequency can be obtained by shunting a trimmer capacitor across the crystal elec-

Crystal Oscillators · 183

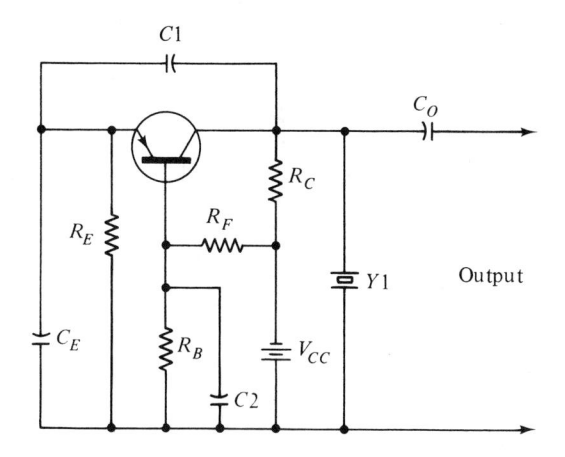

Figure 7-17 Colpitts-type crystal-controlled oscillator arrangement.

trodes. However, this *added capacitance can make the oscillator harder to start, and can contribute to oscillator dropout.* C1 must have a suitable value for rated power output to be realized; moreover, if C1 is off-value, the possiblity of oscillator dropout is aggravated.

Regenerative feedback from collector to base is accomplished by the mutual inductance between transformer windings 1-2 and 3-4. Transformer action provides the necessary 180-deg phase shift for the feedback signal. Resistors R_B, R_F, and R_C provide the base and collector bias. Emitter swamping resistor R_E is bypassed by C_E to maximize the power output. At frequencies above and below the resonant frequency of the crystal, the impedance of the crystal increases and reduces the amount of feedback. This factor prevents oscillation at off-peak frequencies. Note that *a crystal must be ground to operate in a specific circuit (with specified impedance components); the crystal is likely to operate unsatisfactorily, or not at all, in some other arbitrary circuit.*

A Colpitts-type crystal-controlled oscillator arrangement is exemplified in Fig. 7-17. This design employs the parallel-resonant mode of the crystal. In other words, if the crystal were to be replaced by its equivalent circuit shown in Fig. 7-15, its function would become analogous to that of the basic Colpitts arrangement. The circuit in Fig. 7-17 utilizes the CB mode with feedback from collector to emitter via C1. Resistors R_B, R_C, and R_F provide the proper bias and operating conditions for the circuit. Resistor R_E is

184 · Principles of Oscillator Design

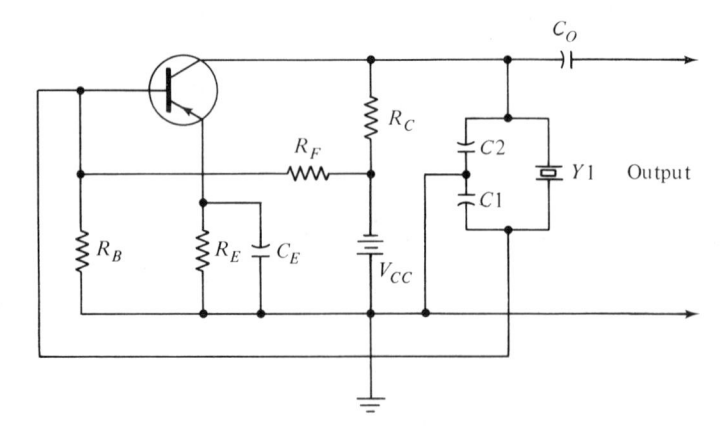

Figure 7–18 A Colpitts-type crystal-controlled oscillator, with collector-base regeneration.

an emitter swamping resistor. Capacitors C1 and C_E form a voltage divider across the output terminals of the oscillator. Capacitor C2 bypasses the signal around the base biasing resistor R_B.

Observe that no phase shift occurs in the foregoing configuration. Accordingly, the feedback signal is connected so that the voltage across C_E will be returned to the emitter with no phase shift. The oscillating frequency is determined not only by the crystal but also by the parallel capacitance presented by C1 and C_E. *Their capacitance value is ordinarily large, in order to swamp both the input and output capacitances of the transistor and to make the operating frequency comparatively independent of variations in transistor parameters.* Since the parallel capacitance of C1 and C_E affects the oscillator frequency, crystal operation is in the inductive region of the impedance-versus-frequency characteristic between the series and parallel resonant frequencies depicted in Fig. 7–15(c).

Next, a Colpitts-type crystal-controlled oscillator with collector-base regeneration is exemplified in Fig. 7–18. The resistors provide suitable operating voltages for the transistor. Capacitors C1 and C2 form the voltage divider for this circuit. Capacitor C_E is the emitter bypass capacitor. In contrast to the foregoing arrangement, a 180-deg phase shift occurs between the input and output signals. Note that the connection between the capacitors is grounded. If a crystal-controlled oscillator is to be used in communications equipment, as in a scanner-monitor radio receiver, it is often desirable to operate the crystal at a *harmonic* or *overtone* frequency. A suitable

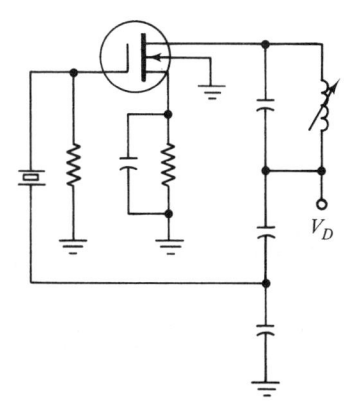

Figure 7-19 Typical crystal-controlled oscillator configuration for overtone operation.

configuration for overtone operation is shown in Fig. 7-19. The oscillator tank is tuned to the third, fifth, or seventh overtone of the crystal. *Note that an overtone crystal must be specially ground.*

An overtone crystal normally oscillates at the selected overtone frequency. Sometimes the crystal will operate at both its fundamental frequency and at its overtone frequency, if the circuit parameters are such that equal Q values prevail at the two frequencies. At "overlapping" frequencies, where either a fifth or a seventh overtone crystal can be used (such as 115 to 125 MHz), the power output from the oscillator will be about the same for either overtone. Note that use of the seventh overtone frequency requires critical circuit adjustment; there is also a potential hazard that a "frequency jump" may occur randomly, with operation occurring for a period at the seventh overtone frequency, followed by a sudden shift to the fifth overtone frequency, and then back to the seventh overtone frequency. Therefore, *the circuit designer must make a careful worst-case analysis of overtone oscillator design.* This is ordinarily done on an experimental basis.

CHAPTER 8

Nonsinusoidal Oscillator Circuit Design Principles

8–1 GENERAL CONSIDERATIONS

Free-running (astable) multivibrators comprise the most basic class of nonsinusoidal oscillators. A multivibrator is a type of relaxation oscillator that generates rectangular waveforms. Both square waves and pulses are included in this category. A basic astable multivibrator arrangement is shown in Fig. 8-1. It is essentially a two-stage direct-coupled amplifier with positive feedback from output to input. One stage conducts and the other stage is cut off for a pre-determined time, after which their states suddenly reverse. The period of oscillation is based on the time constants $R_{F1}C_{F1}$ and $R_{F2}C_{F2}$. In this example, the time constants are equal, and a square waveform is developed at the collector of each transistor. Note that if these time constants were unequal, a rectangular waveform would be developed at the collector of each transistor.

If a square-wave output is desired from the configuration in Fig. 8-1, *the worst-case condition occurs when one RC coupler has its positive-limit tolerances and the other coupler has its negative-limit tolerances.* For example, suppose that C_{F1} and R_{F1} are each 20 percent high, and that C_{F2} and R_{F2} are each 20 percent low. In this situation, there will be a difference of more than two to one in the periods of successive half-cycle outputs. Note that C_{E1} and C_{E2} bypass the emitter swamping resistors, and have no effect on the output waveform or frequency of operation. Since the output waveform

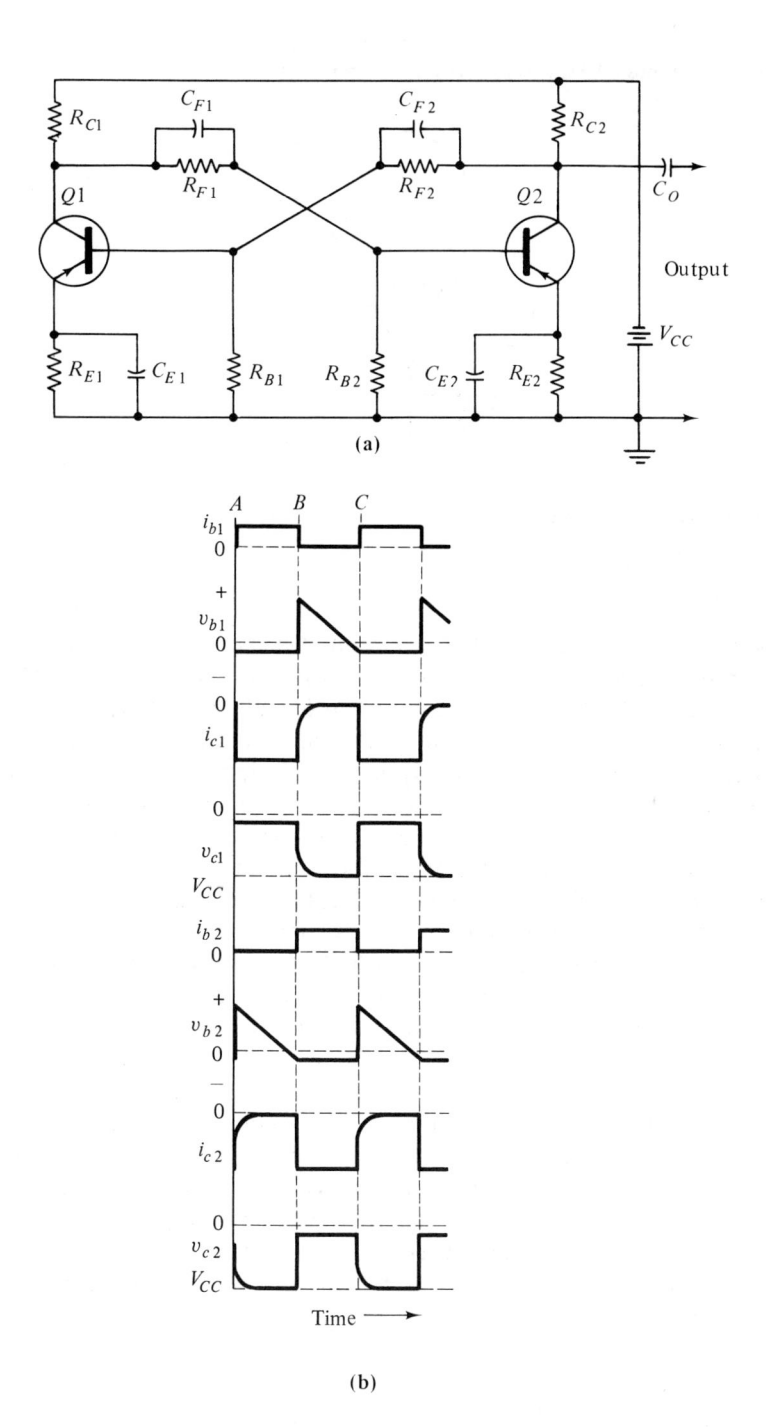

Figure 8–1 Basic astable multivibrator arrangement: **(a)** configuration; **(b)** operating waveforms.

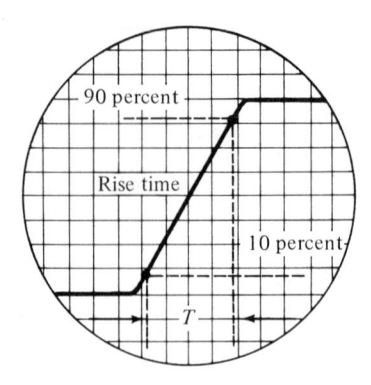

Figure 8-2 Rise time T is measured from 10 percent to 90 percent of maximum waveform amplitude.

is not entirely flat-topped, the circuit designer may feed the multivibrator output into a clipper unit in order to improve the waveform.

Multivibrators can be grouped into *high-current* types and *low-current* types. A high-current design employs emitter swamping resistors that have approximately half the value of the collector load resistors. In turn, i_{c1} and i_{c2} have comparatively high values. This design is reasonably tolerant of supply-voltage variation. Note that the value of R_{F1} is typically ten times the value of R_{C1}. The amplitude of the output waveform is somewhat greater than half the value of V_{cc}. To calculate the maximum power dissipation of each transistor, it is good practice to utilize *conservative design* procedure, and to assume that the full supply voltage is applied to the transistor with current limited only by the series resistance in the circuit. For example, if R_{C1} were 1000 ohms, and R_{E1} were 500 ohms, and V_{cc} were 10 V, a maximum collector-current flow of approximately 7 mA would be assumed. If it is further assumed that the collector-emitter potential has a maximum value of 10 V, the transistor has a maximum power dissipation of 70 mW. This conservative design procedure allows reasonable margin for supply-voltage variation, and transistor derating for increased operating temperature.

A transistor in a multivibrator circuit operates as a switch, and *worst-case design is based on the premise that overdrive will demand three times the base current supplied by the DC bias network.* Note that overdrive is required to obtain a fast rise time (see Fig. 8-2). Fundamentally, the rise time is limited by the beta cutoff frequency of the transistor. Rise time is further limited by the interval required for the transistor to come out of saturation (the arrangement in Fig. 8–1 operates in the saturated mode). This limitation is taken into account in the rated switching time of the transistor. Since a high-current multivibrator design is considered in this example, the values of the collector load resistors are comparatively small, and do not enter prominently into rise-time parameters. A typical switching-type transistor has the following switching-time ratings:

Delay time, 0.2 μs
Rise time, 1 μs
Storage time, 1 μs
Fall time, 1.2 μs

These ratings apply to the transistor only (bogie values) and do not take circuit characteristics into account. Thus, when the transistor is cut off and is then suddenly driven into conduction, it will require 0.2 μs before the change in base voltage is reflected as a change in collector voltage. Then, it will require 1 μs for the leading edge of the output waveform to rise from 10 percent to 90 percent of its maximum amplitude. Next, with the transistor in saturation, when the base is suddenly driven into cutoff, storage of charge carriers in the base region will cause the output amplitude to remain at almost its maximum amplitude for 1 μs. After this storage time has passed, the output waveform starts to fall, and it will require 1.2 μs for the trailing edge of the waveform to fall from 90 percent to 10 percent of its total excursion.

Calculation of bias-circuit values is made on the basis of amplifier operation. The circuit designer assumes that there is no feedback in the circuit, and that both transistors are to be forward-biased so that the collector will have a potential equal to 0.6 of the supply voltage, and the emitter potential will be equal to approximately 0.2 of the supply voltage. The values of base resistance (R_{B1} and R_{B2} in Fig. 8–1) determine the base voltage. *To make a bias test, the circuit designer should breadboard the circuit with the coupling capacitors omitted.* In turn, the theoretical frequency of oscillation greatly exceeds the beta cutoff frequency of the transistors, and the circuit is stable. Accordingly, the designer can measure the transistor terminal voltages, and adjust the base-resistance values as required. *Worst-case bias relations can be checked concomitantly.*

Low-current Multivibrator   A low-current multivibrator utilizes collector resistors that have approximately ten times the value of the emitter swamping resistors. In turn, i_{c1} and i_{c2} have comparatively low values. This design is significantly responsive to changes in supply-voltage value; an increase in supply voltage causes an increase in the oscillating frequency. In some applications, this characteristic is desirable. For example, a low-current multivibrator may be employed as a voltage-to-frequency converter to transmit voltage information in a telemetry system. Bias tests and worst-case bias relations can be checked in the same manner as explained previously for the high-current type of multivibrator.

8-2 BISTABLE MULTIVIBRATOR OPERATION

A bistable multivibrator is also called a *flip-flop*. Its circuitry is similar to that of an astable multivibrator, except that the bias circuitry prevents a nonconducting transistor to come out of cutoff until a trigger pulse is applied. Thus, a bistable multivibrator is initially at rest in one of its two stable states. When triggered by an input pulse, the multivibrator switches to its second stable state, where it remains until triggered by another pulse. Two basic classes of bistable multivibrators, or flip-flops, are termed the *saturating* and the *nonsaturating* types. A typical saturating type of flip-flop is shown in Fig. 8-3. Each transistor is held in its conducting or nonconducting mode by the other transistor. Bias relations are exemplified in Fig. 8-4. The essential consideration is that tolerances on components and devices cannot result in a transistor coming out of cutoff in the absence of a trigger pulse. *Worst-case analysis is made on the basis of these voltage-divider conditions.*

When one of the transistors in a flip-flop is cut off, its output resistance is high and its collector current is effectively zero. However, current flow through the voltage-divider network (R_L, R_F, and R_B) owing to collector-battery voltage V_{cc} (−28 volts) and base-battery voltage V_{bb} (+2 volts), causes resultant voltage drops across each resistor. In this example, the voltage drop across R_L is 2 volts, leaving approximately the full value of V_{cc} at point C. Voltage drops across R_F and R_B are 21 volts and 7 volts, respectively. This provides a negative voltage (−5 volts) at point B, which forward-biases Q2. In turn, Q2 will rest in its conducting state. At the same time, Q1 will rest in its cutoff state, as explained next.

High collector current (saturation) of the *on* transistor flows through R_L and causes a voltage drop equal to V_{cc}, so that the volt-

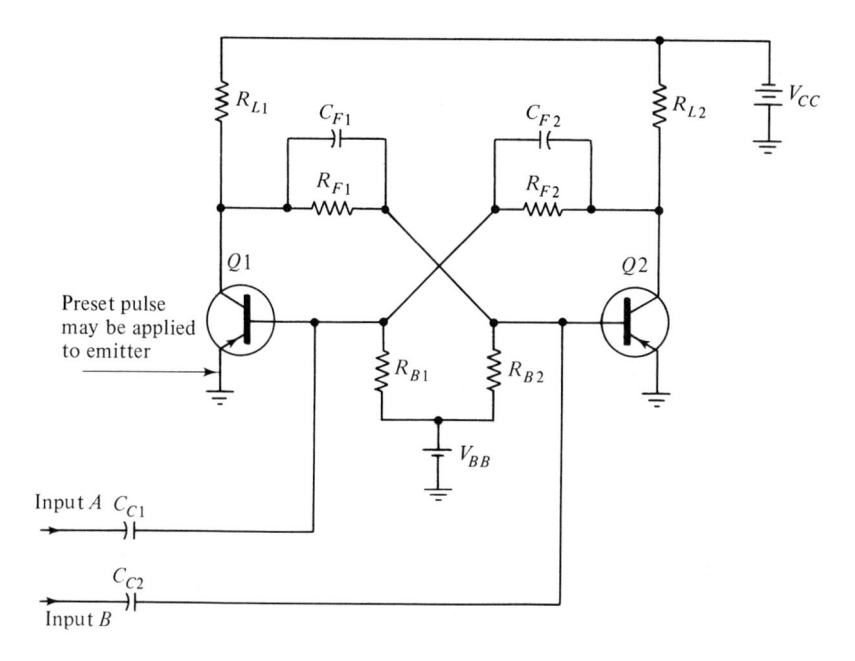

Figure 8-3 Conventional bistable multivibrator configuration.

age at point C is effectively zero, as depicted in Fig. 8-4(b). Division of the base-battery voltage V_{bb} (+2 volts) by R_F and R_B results in a positive voltage (+1.5 volts) at point B. Application of a negative pulse to the base of the *off* transistor, or a positive trigger pulse to the base of the *on* transistor will switch the conducting state of the circuit, provided that the trigger pulse has sufficient amplitude to cross the threshold. Accordingly, *the circuit designer must take tolerances on the amplitude of the trigger pulse into account in a worst-case analysis.* Note that collector triggering may be employed in lieu of base triggering. In either case, during the rapid transition period, *both* transistors conduct. That is, conduction is increasing in one transistor but is decreasing in the other transistor. Speed of transition is facilitated by provision of the small bypass (speed-up) capacitors C_{F1} and C_{F2} in Fig. 8-3.

Trigger Pulse Steering Several methods can be utilized for application of a trigger pulse or pulses to a flip-flop. Basically, the best method is determined by the polarity and amplitude of the specified trigger pulse, and the desired repetition rate. In most cases, a very

192 · Nonsinusoidal Oscillator Circuit Design Principles

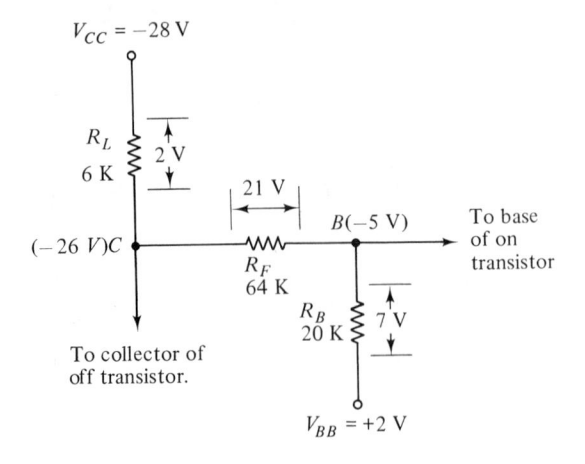

$V_{CC} = -28$ V

R_L 6 K, 2 V

21 V

$B(-5$ V)

$(-26\ V)C$

R_F 64 K

R_B 20 K, 7 V

To base of on transistor

To collector of off transistor.

$V_{BB} = +2$ V

(a) Off (cutoff) transistor.

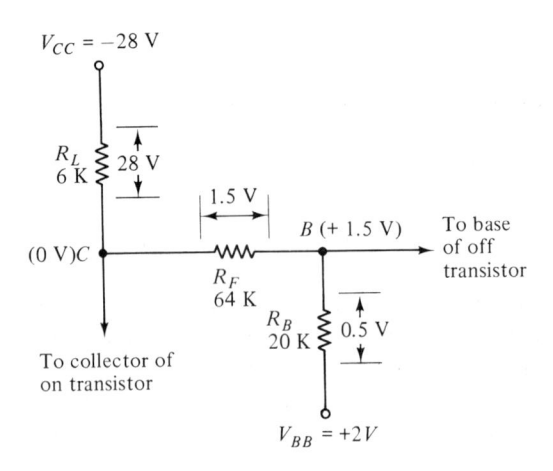

$V_{CC} = -28$ V

R_L 6 K, 28 V

1.5 V

$B\ (+1.5$ V)

$(0\ V)C$

R_F 64 K

R_B 20 K, 0.5 V

To base of off transistor

To collector of on transistor

$V_{BB} = +2V$

(b) On (saturated) transistor

Figure 8–4 Voltage-divider networks to show the saturated and cutoff conditions of a bistable multivibrator.

narrow trigger pulse is used, or a step voltage may be applied. Either will cause the same circuit reaction. If PNP transistors are used, a negative trigger pulse may be applied to the base of the *off* transistor and thereby drive the transistor into conduction, or a positive trigger pulse may be applied to the base of the *on* transistor and thereby drive it out of saturation and into the transition region. Again, a negative trigger pulse may be applied to the collector of the *on* transistor. Conversely, a positive trigger pulse may be applied to the collector of the *off* transistor.

Because a flip-flop cannot distinguish between a normal trigger pulse and a noise pulse or a spurious "glitch" pulse, *the circuit designer must take every possible precaution against the introduction of switching transients or any other form of stray pulses that could cause false triggering.* In certain applications, it is desirable to trigger a bistable multivibrator on and off repeatedly with trigger pulses of the same polarity. If the pulse were applied simultaneously to both the *off* and to the *on* transistors, *switching time would be delayed.* On the other hand, with the addition of pulse-steering diodes, an input trigger pulse is directed to the proper transistor only, and this pulse will drive the intended transistor *off* or *on*. A bistable multivibrator configuration with negative-pulse steering diodes is shown in Fig. 8-5. Inspection of this arrangement reveals that it automatically provides optimum trigger action. A bistable multivibrator arrangement that employs positive-pulse steering is depicted in Fig. 8-6.

Assume that a negative pulse is applied to the bases of both input transistors from the same source in the arrangment of Fig. 8-3. The tendency is to drive the *off* transistor into conduction and to drive the *on* transistor into saturation. Thus, the feedthrough from the collector of the *off* transistor must overcome the initial bias condition of the *on* transistor and also the additional bias provided by the input pulse. Accordingly, the circuit designer would have to provide a comparatively large trigger amplitude, and the switching time would be slowed down. In other words, *this mode of operation develops slower rise and fall times.*

Next, negative or positive pulse steering may be provided in the multivibrator arrangement, as depicted in Fig. 8-7. In PNP-type transistors, a negative trigger pulse drives the *off* transistor *on;* a positive trigger pulse drives the *on* transistor *off*. Polarities are reversed for NPN-type transistors. In each of these steering methods, the base of the *on* transistor is negative, and the base of the *off* transistor is positive with no applied signal. *It is good design practice to avoid loose tolerances on the steering diodes, although this is not a highly critical application that requires matched pairs.*

For negative pulse-steering operation, diodes CR1 and CR2 are arranged as shown in Fig. 8-7(a). Resistors R1 and R2 form a

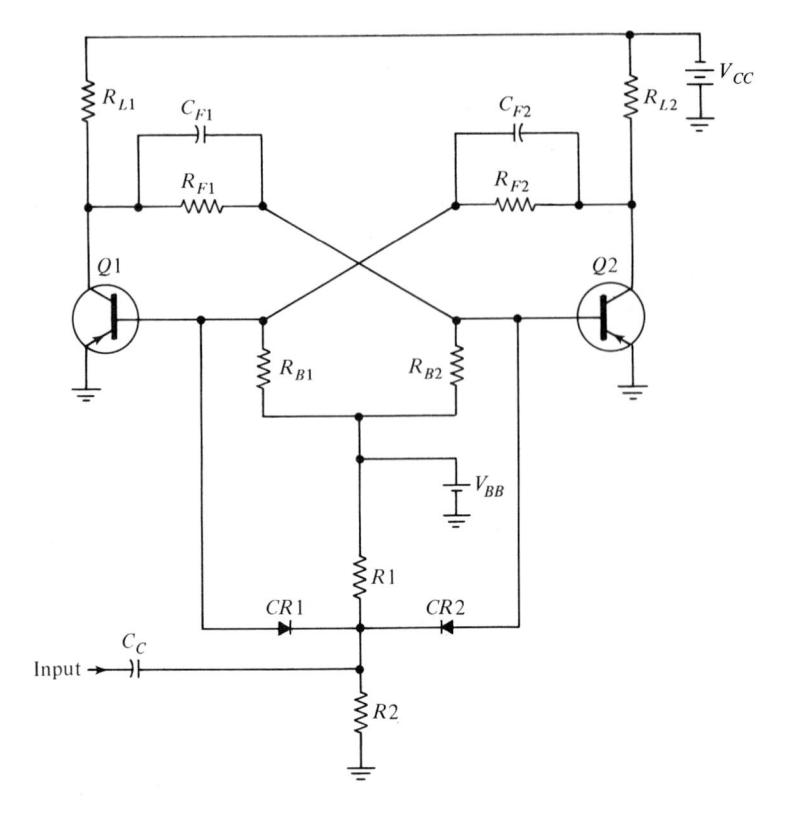

Figure 8–5 Bistable multivibrator with negative-pulse steering diodes.

voltage divider so that the cathodes of the diodes are at a positive potential. Diode CR1 is in a nonconducting state because of reverse bias, and diode CR2 is in a conducting state owing to forward bias. When a negative trigger pulse is applied through coupling capacitor C_c, the reverse-bias diode CR1 prevents the pulse from being applied to the base of the *on* transistor. Diode CR2 is forward-biased, and the negative trigger pulse is applied to the base of the *off* transistor, thereby overcoming the positive bias on the base and driving the transistor *on*. In turn, the multivibrator changes state. Next, a second trigger pulse is directed through diode CR1, which would then be biased in the forward direction. Note that a positive trigger pulse applied to the steering circuit, regardless of the state of the multivibrator, increases the reverse bias on the diodes and therefore has no effect on the state of the configuration.

For positive pulse steering, depicted in Fig. 8–7(b), diode po-

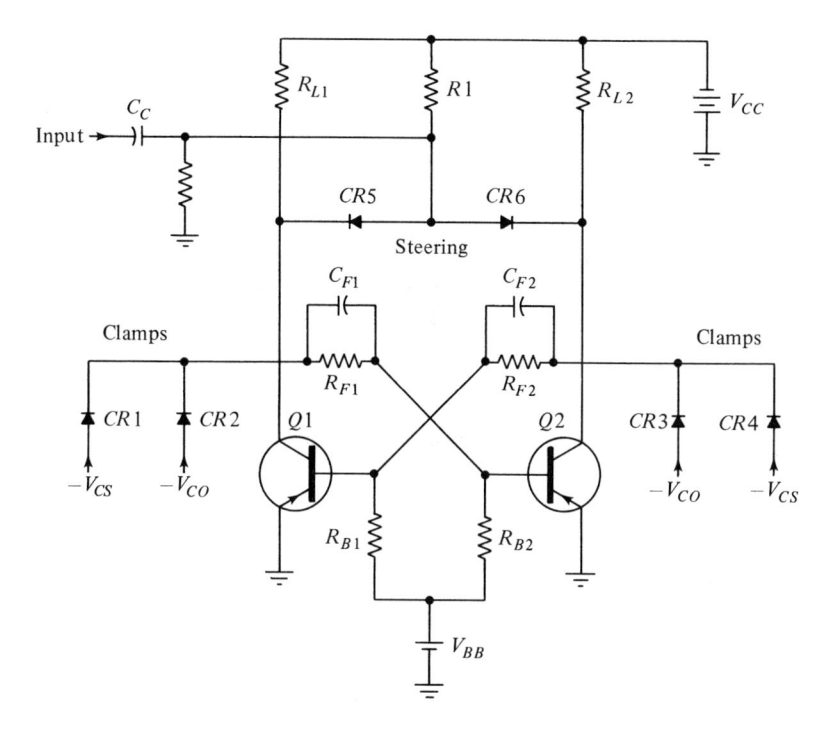

Figure 8–6 A nonsaturating bistable multivibrator configuration with diode clamping and positive-pulse steering.

larities are reversed and a negative potential is provided at the junction of the voltage divider that consists of resistors R1 and R2. Diode CR1 is forward-biased, whereas diode CR2 is reverse-biased. A positive trigger pulse, accordingly, is applied through CR1 to the base of the *on* transistor, thereby driving it *off*. The reverse bias on diode CR2 prevents the trigger pulse from being applied to the base of the *off* transistor. As the circuit switches state, bias conditions on the diodes are reversed, and a second trigger pulse is directed through the forward-biased diode CR2 to the base of the *on* transistor. Note that a negative trigger pulse applied to this arrangment increases the reverse bias on the diodes and therefore has no effect on the circuit.

Note that, when power is applied to the arrangement shown in Fig. 8–3, the circuit may start operation either with Q1 *on,* or with Q2 *on*. In many applications, the start-up mode is indifferent. On the other hand, in other applications, such as binary counters, the start-up mode is a prime aspect of circuit action. *In such a case, the cir-*

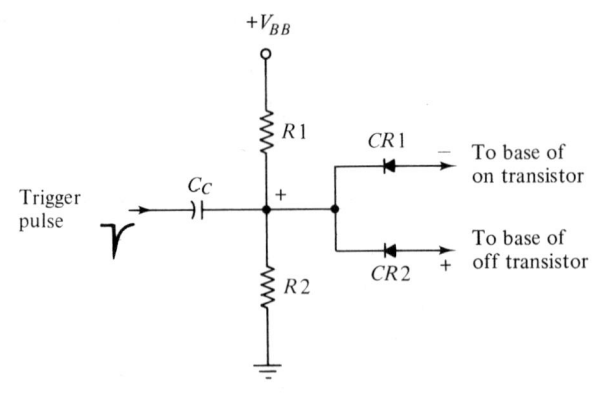

196 · Nonsinusoidal Oscillator Circuit Design Principles

(a) Negative pulse steering circuit

(b) Positive pulse steering circuit

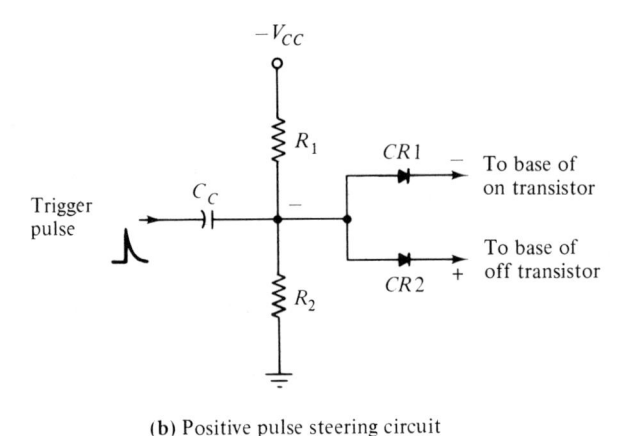

Figure 8–7 Pulse-steering diode arrangements.

cuit designer elaborates the basic flip-flop configuration with a preset line. For example, a preset positive pulse may be coupled to the emitter of Q1, to ensure that this transistor is turned on after start-up, or to reverse the existing *off* state of the transistor at any time in an operating sequence. Conversely, a negative preset pulse coupled to the emitter of Q1 will ensure that this transistor is turned off at any chosen time.

8-3 NONSATURATING BISTABLE MULTIVIBRATOR

In the nonsaturating type of bistable multivibrator, as exemplified in Fig. 8-6, lower and upper limits are established for the output voltage in the quiescent states. By thus avoiding transistor saturation, circuit response time is decreased and the configuration can be triggered at a higher repetition rate. Diodes CR1 and CR2, biased by V_{CS} and V_{CO}, provide *saturation and cutoff clamping* for Q1. Diodes CR3 and CR4, also biased by batteries V_{CS} and V_{CO}, provide saturation and cutoff clamping for Q2. These clamp diodes effectively short-circuit overdrive voltages to ground that would otherwise drive the transistors into saturation. Positive pulse-steering diodes are included in the trigger circuitry of the configuration. *The circuit designer must take the tolerances on both the clamping diodes and the transistors into account in evaluating the worst-case condition.* This is usually accomplished on an experimental basis.

8-4 MONOSTABLE MULTIVIBRATOR

A monostable multivibrator, also called a one-shot multivibrator, is a hybrid design that includes features of both the astable and bistable arrangements. A typical configuration is shown in Fig. 8-8 with operating waveforms. Bias voltages and regenerative action hold Q2 in saturation and Q1 at cutoff during the quiescent period. If a narrow pulse is applied, the monostable multivibrator proceeds through a cycle of operation. Its ouptut waveform is the same, regardless of the shape of the trigger pulse. Battery V_{cc} provides the necessary collector potentials and the forward bias for Q2. Transistor Q2 is in saturation during the quiescent period. Negative potential v_{c2} at the collector of Q2 is effectively zero, or ground potential, owing to the transistor's saturated condition.

The reverse bias provided by battery V_{BB} and the voltage developed across R_{B1} maintains Q1 at cutoff. Collector potential v_{c1} is negative and is equal to the value of battery voltage V_{CC}. Capacitor C_{F1} permits rapid application of the regenerative signal from the collector of Q1 to the base of Q2, and it is charged to the battery voltage V_{CC} through R_{L1}, and the essentially short-circuited base-emitter junction of forward-biased transistor Q2. Various means of triggering may be utilized; here, let us consider the application of a negative pulse at the input. A negative pulse applied through coupling capacitor C_C to the base of Q1 causes the transistor to start

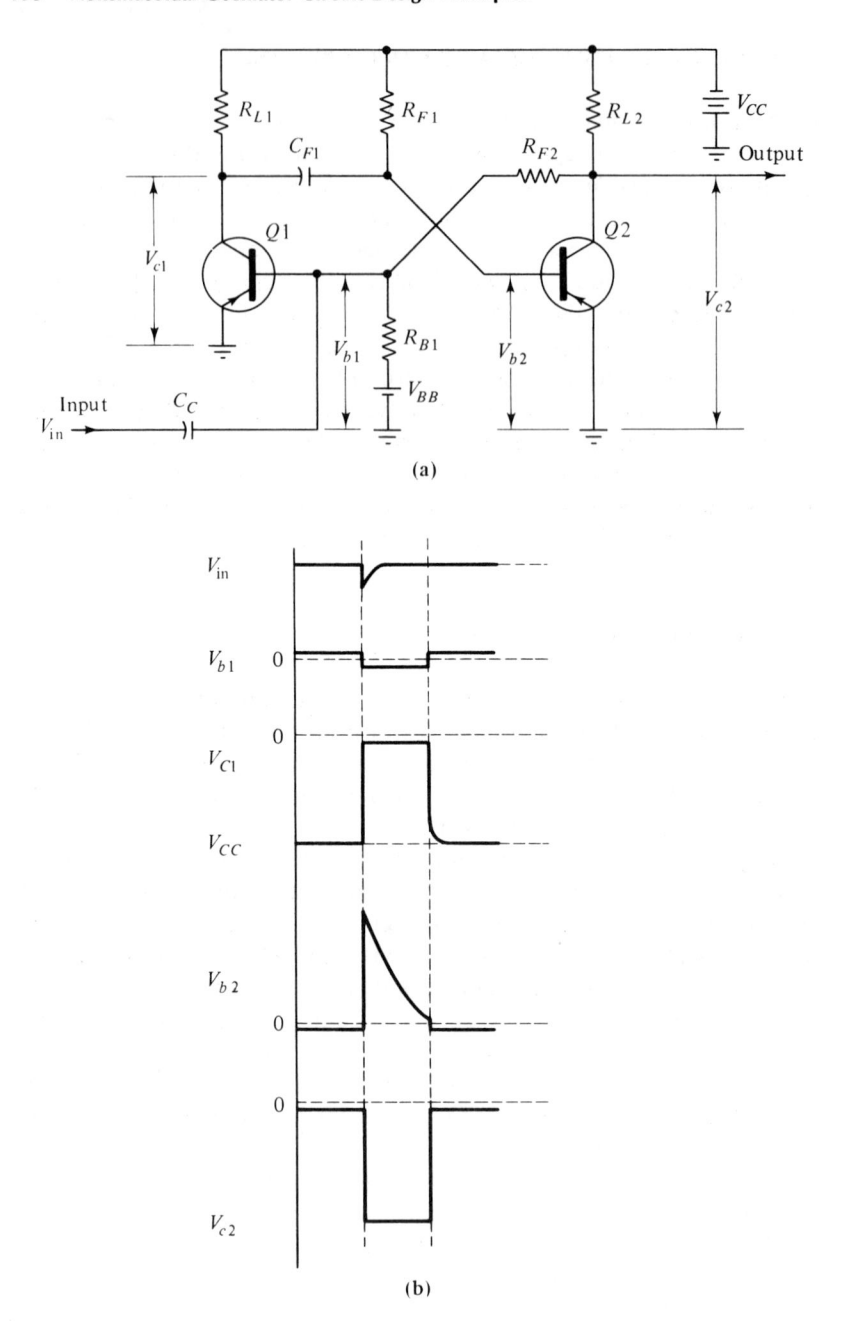

198 · Nonsinusoidal Oscillator Circuit Design Principles

Figure 8-8 Monostable multivibrator arrangement: **(a)** configuration; **(b)** operating waveforms.

conduction. The high negative voltage at the collector of Q1 starts to fall (becomes more positive). This positive-going voltage is coupled to the base of Q2, and decreases its forward bias. Base and collector current of Q2 decrease, and the collector voltage of Q2 increases negatively. A portion of this voltage is coupled through R_{F2} to the base of Q1, and increases its negative potential.

The foregoing regenerative action rapidly drives Q1 into saturation and Q2 into cutoff. Since C_F was initially charged almost up to V_{CC}, the base of Q2 is at a positive potential almost equal to V_{CC}. Capacitor C_{F1} discharges through R_{F1} and the low saturation resistance of Q1. Base potential v_{b2} of Q2 becomes less positive, and Q2 again conducts as its base becomes slightly negative. The collector voltage on Q2 increases positively and is coupled to the base of Q1, driving the transistor into cutoff. Transistor Q1 is again at cutoff and Q2 is in saturation with its collector voltage almost zero. This is the stable condition until the next trigger pulse is applied. The pulse output is taken from the collector of Q2. Its width is determined primarily by the time constant $R_{F1}C_{F1}$.

All device and component tolerances must be thoroughly checked out in a worst-case analysis. If a regulated power supply is not utilized, fluctuations in supply voltages should also be taken into account. Temperature variation can shift the trigger level, or defeat it completely. *Reasonable deterioration in transistor characteristics owing to aging factors should be observed.* Tolerance on the trigger-pulse amplitude can be an important consideration. The trigger pulse train should be free from "glitches," high-amplitude noise pulses, or switching transients that could cause false triggering.

8–5 SCHMITT TRIGGER (SQUARING) CIRCUIT OPERATION

An important variation on the conventional bistable multivibrator configuration, called the Schmitt trigger circuit, is illustrated in Fig. 8–9. This arrangement produces either a square-wave or a rectangular-wave output, depending upon the waveshape of the input signal. *A sine-wave input results in a square-wave output. A nonsinusoidal input that has unsymmetrical positive and negative excursions results in a rectangular-wave output.* For this reason, the Schmitt trigger circuit is called a squaring circuit. It differs from the conventional bistable multivibrator arrangement in that one of the coupling networks is replaced by a common-emitter resistor. This resistor provides emitter coupling between Q1 and Q2 and additional regeneration. The added regeneration provides increased switching rapidity.

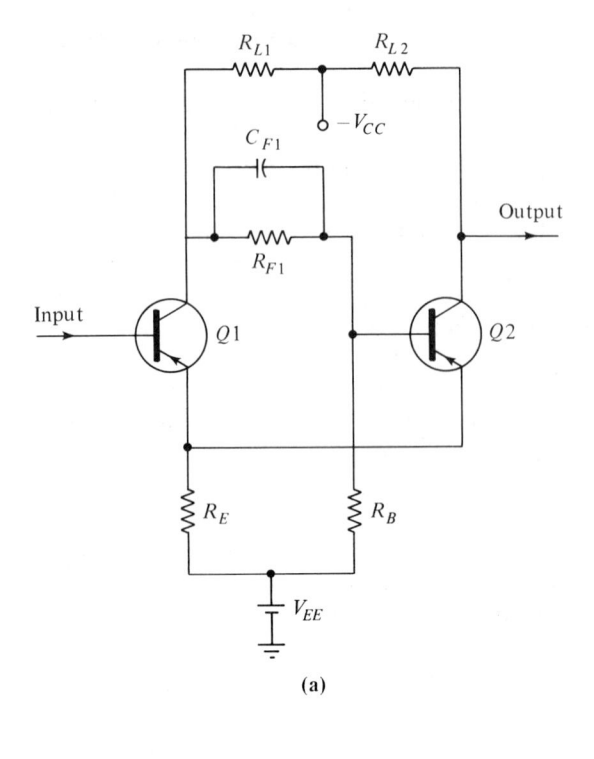

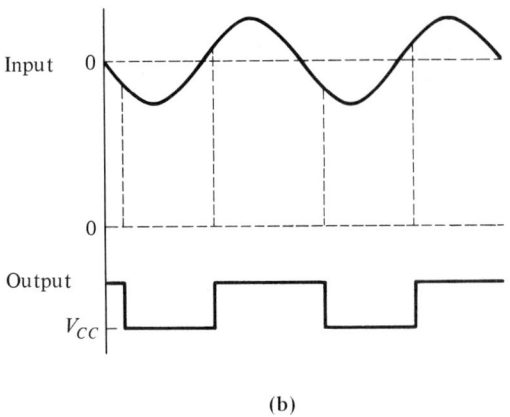

Figure 8-9 Schmitt trigger (squaring) circuit: **(a)** Configuration: **(b)** Operating waveforms.

If we assume a quiescent condition with no input (Fig. 8-8) and with Q1 cut off, the collector voltage is equal to V_{CC}. This negative voltage is coupled to the base of Q2 via R_{F1}, and the base voltage of Q2 is equal to the voltage drop across R_B minus the battery voltage V_{EE}. Current flow from the emitter of Q2 through R_E to bat-

200

tery V_{EE} maintains the emitter of Q1 at a negative potential. The reverse bias now developed between the emitter and base of Q1 maintains the cutoff condition. The high negative bias at the base of Q2 provides forward bias for the base-emitter junction and causes the transistor to rest in its saturation region.

A negative-going signal of sufficient amplitude applied to the base of Q1 will overcome the reverse bias and cause Q1 to conduct. Its collector potential becomes less negative, and this change is coupled to the base of Q2. The emitter current of Q2 decreases, thereby lowering the potential across R_E. In turn, the emitter of Q1 becomes less negative, thereby reducing the reverse bias and increasing the collector current. This regenerative action continues until Q1 is saturated and Q2 is cut off. Now, the output voltage is at its maximum negative value. This stable condition continues until the input waveform becomes positive-going. Then the base potential of Q1 decreases and the reverse bias increases; the collector voltage becomes more negative, the emitter current decreases, and the potential across R_E decreases. Simultaneously, the increasing negative voltage at the collector of Q1 is coupled to the base of Q2 and drives it negative; the decreasing voltage across R_E causes the emitter of Q2 to go more positive. Both of these actions reduce the reverse bias of the emitter-base junction, and Q2 again operates in saturation, thereby cutting off Q1 and returning the circuit to its original operating state.

It follows from this sequence of circuit actions that a sine-wave input will not produce a square-wave output unless the two transistors are reasonably well matched in characteristics. In other words, unequal threshold levels will result in a rectangular-wave output. Note that the Schmitt trigger circuit is extensively used as a *voltage-level detector* in both analog and analog-digital systems. In digital-system applications, only "high" and "low" levels need be sensed. On the other hand, in analog-system applications, precise voltage levels may need to be sensed. *The required precision must be taken into careful consideration by the circuit designer in his evaluation of worst-case conditions.* When very precise voltage levels are to be sensed, it is preferable to employ an op-amp comparator arrangement instead of a Schmitt trigger circuit.

8–6 TRIGGERED BLOCKING OSCILLATOR OPERATION

A basic triggered blocking-oscillator (monostable type) and a free-running blocking oscillator are similar except for bias arrangements.

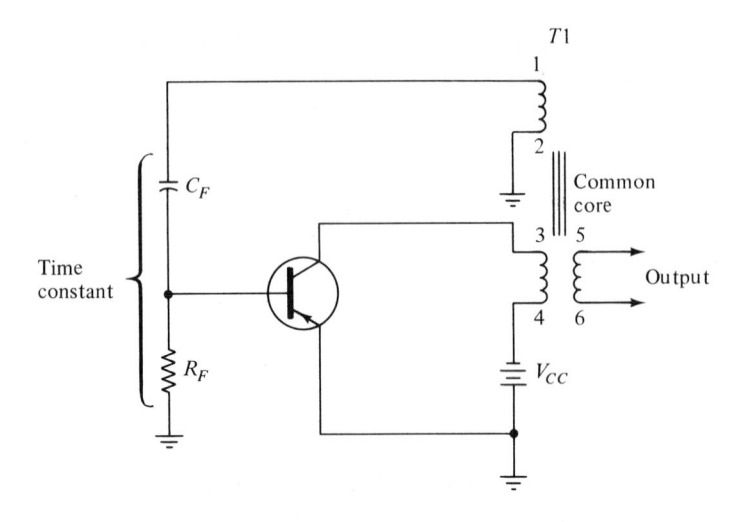

Figure 8–10 A basic blocking-oscillator configuration.

A free-running blocking oscillator may be synchronized by injection of pulses that have a slightly faster repetition rate than the oscillator's free-running frequency. A basic configuration for a free-running blocking oscillator is shown in Fig. 8–10. When the power is applied to the circuit, current rises rapidly in the base branch owing to forward bias established between base and emitter by V_{CC}. Current flow in the collector circuit increases in accordance with this base driving current. Because of increasing collector current, a voltage that becomes more negative is induced in the transformer winding 1–2. This voltage charges capacitor C_F through the small forward resistance of the base-emitter junction, and appears across this resistance to increase the forward bias voltage. Regeneration continues rapidly until the transistor is saturated.

As the transistor goes into saturation, its collector current becomes constant. Therefore, there is no longer any induced voltage in winding 1–2 and a charging potential is no longer applied across C_F. In turn, capacitor C_F begins discharging through R_F. The field around winding 1–2 collapses and induces a voltage of reverse polarity in the winding (counter emf). That is, the base is driven positive (reverse bias). Base and collector current fall to zero. The transistor is held at cutoff until C_F, discharging through R_F, reaches the point at which the transistor is forward-biased and conduction ensues. The operating cycle is then repeated.

The output waveform from a blocking oscillator is a pulse, the width of which is determined primarily by the inductance and distributed capacitance of winding 1-2. The time between output pulses (resting or blocking time) is determined by the time constant $R_F C_F$. Output voltage is coupled through transformer winding 3-4 to the load by winding 5-6. Next, observe the configuration for a triggered blocking oscillator, exemplified in Fig. 8-11. In this monostable arrangement, the base-emitter junction of the transistor is reverse-biased, and the transistor is held at cutoff. In turn, an output consisting of a single pulse is obtained each time that the input circuit is triggered. As shown in the output waveform example, the output pulse is accompanied by substantial backswing. The transistor is also subject to minority charge-carrier storage. *These characteristics can be avoided by the circuit designer at the expense of moderate elaboration, as explained below.*

With reference to Fig. 8-11, the transistor is held at cutoff in its quiescent state. The base-emitter junction is reverse-biased by V_{BB}. Also, the collector-base junction is reverse-biased by V_{CC}. Transformer T1 provides regenerative feedback and couples the output signal to the load. Capacitor C_F functions as a DC blocking capacitor. Capacitor C_C couples the input signal to the base of the transistor. When a negative trigger pulse is applied at the input, the transistor starts to conduct. Collector current flows through winding 3-4 and produces a varying magnetic flux that induces a voltage of opposite polarity into winding 1-2. This voltage is coupled via C_F to the base of the transistor and supplies negative feedback. Regenerative action continues until the transistor is driven into saturation and the collector current ceases to increase. Since the collector current becomes constant, no feedback voltage is induced in T1, and the reverse bias provided by V_{BB} cuts off the transistor. Collector current continues to flow briefly because of *minority charge-carrier storage.*

When collector current stops flowing, the collapsing magnetic field around winding 1-2 induces a voltage in winding 3-4 that exceeds the value of battery voltage V_{CC}. In some instances, *this backswing voltage may exceed the collector breakdown voltage rating of the transistor.* Driving the transistor into its saturation region introduces the undesired condition of minority charge-carrier storage. A means of avoiding both of the foregoing disadvantages is realized in supplementary diode clamping facilities. Observe the arrangement depicted in Fig. **8-11C**. Inclusion of clamping diodes CR1 and CR2 prevents transistor saturation and also avoids excessive induced voltage at the collector of the transistor. *These circuit actions increase the maximum pulse repetition rate that can be obtained, and also make transistor replacement less critical.*

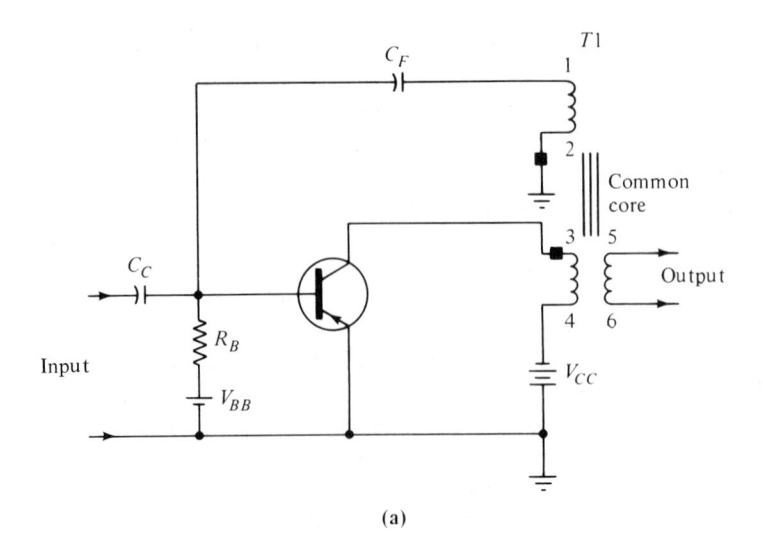

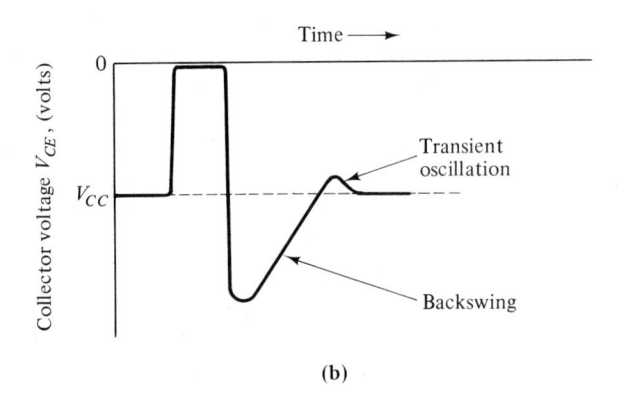

204 · Nonsinusoidal Oscillator Circuit Design Principles

Figure 8–11 Example of a triggered blocking-oscillator arrangement. **(a)** Configuration. **(b)** Output waveform.

Note that the base-emitter junction of the transistor is reverse-biased by V_{BB}. The collector is reverse-biased by the combined voltages of V_{C1} and V_{C2}. Quiescently, diode CR2 has no applied bias and diode CR1 is reverse-biased. When a negative trigger pulse is applied, the transistor is driven from cutoff into conduction. Collector current increases rapidly as a result of circuit regeneration. In turn, the collector voltage falls from the combined value of V_{C1} and V_{C2} to the value of V_{C1}. As the collector potential decreases owing to in-

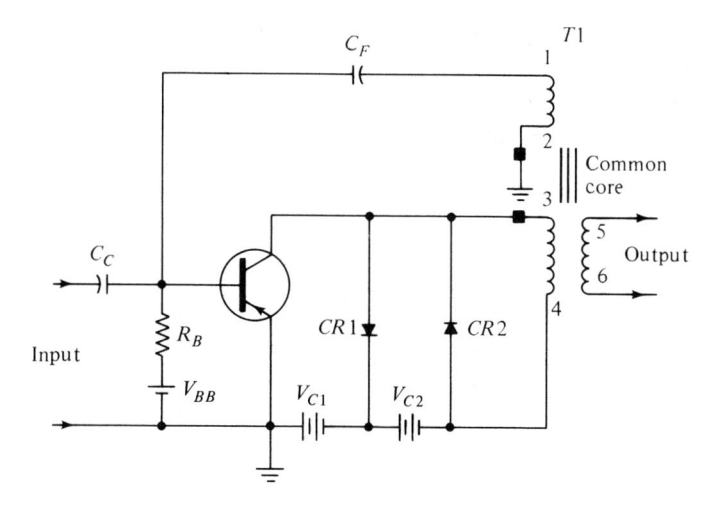

Figure 8-11C Typical triggered blocking-oscillator configuration with clamping diodes.

crease in collector current, the voltage across CR2 develops a reverse bias. Therefore, the effect of CR2 at this time may be neglected. The reverse-bias potential applied to CR1 decreases until the collector voltage reaches a value equal to V_{C1}. Further increase of collector current causes a further decrease in collector voltage, which results in CR1's becoming forward-biased.

At this point, CR1 conducts and carries away current from the collector to maintain collector voltage at the value of V_{C1}. Diode CR1 functions as a short circuit for transformer winding 3-4. Thus, the transistor is prevented from going into saturation. When the magnetic field starts to collapse, the collector voltage becomes more negative, and the bias voltage across CR1 reverts to its nonconducting state when the value of V_{C1} is exceeded. The collector voltage increases negatively until it reaches the value of the combined voltages of V_{C1} and V_{C2}. Any attempt by the induced voltage to further increase the negative voltage at the collector is overcome by conduction of CR2. The stored energy from winding 3-4 is then discharged through this diode, thus preventing the application of excessive voltage between collector and emitter of the transistor. The negative-going voltage does not drop below voltage V_{CC}, and the voltage swing below V_{CC} that occurs in an unclamped circuit is eliminated by conduction of CR2.

8-7 SWITCHING-CIRCUIT CHARACTERISTICS

Nonsinusoidal oscillators exploit switching-circuit characteristics. State transitions are normally characterized by large-signal, or *nonlinear* operation of transistors. In large-signal operation, or nonlinear operation, a transistor functions as an overdriven amplifier with resultant changes in the conduction state and the DC characteristics of the transistor. Advantages and disadvantages of employing different transistor configurations in switching circuits will be considered from the viewpoint of the circuit designer. With respect to *output characteristics,* the CE output from a typical PNP transistor is depicted in Fig. 8-12(a). The characteristics are arranged in three regions: *cutoff, active,* and *saturation*. An arbitrarily chosen load line and the maximum power dissipation curve are also shown.

Cutoff and saturation regions in Fig. 8-12 are considered the *stable* or quiescent regions of operation. A transistor is considered to be in the *off* state (nonconducting) or to be in the *on* state (conducting), when it is operated in the cutoff or saturation region, respectively. The third region of operation, termed the active region, is regarded as the *unstable* (transient) region through which the transistor's operating point passes while changing from the *off* state to the *on* state, or vice versa. The effect of base bias voltage V_{BE} on the collector current I_C is illustrated in Fig. 8-12(b). A typical transistor switching circuit is shown in Fig. 8-13(a). Switch S1 controls the polarity and value of base current from V_{B1} or V_{B2}. Resistors R_{B1} and R_{B2} are the current-limiting resistors. The emitter-base and collector-base diode and switch equivalent circuits representing the *off* and *on* (DC) conditions of the transistor switching circuit are shown in Fig. 8-13(b) and (c).

Note in Fig. 8-12(a) that the load line is shown passing through the area beyond the maximum power dissipation curve. For *normal amplifier operation* of a transistor, this would be very poor practice. In switching circuits, however, the excursion of the collector current through this area is very rapid, and the *average power dissipation* within the transistor falls within the acceptable maximum limitation. The cutoff region includes the area above the zero base current curve ($I_B = 0$). Ideally, with no initial base current, there would be zero collector current; the collector potential would equal V_{CC}. However, at point X on the load line, a small amount of collector current is measured. This is more clearly indicated in Fig. 8-12(b), where at zero base bias voltage (point X), collector current I_C equals approximately 0.05 mA. This is the reverse bias collector

Switching-Circuit Characteristics · 207

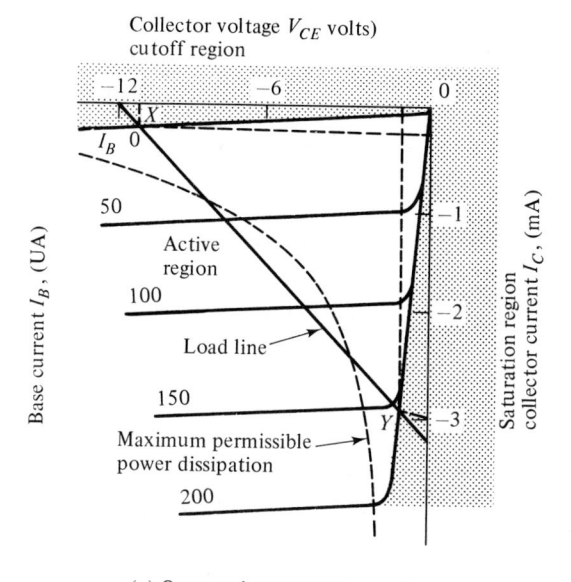

(a) Output characteristics

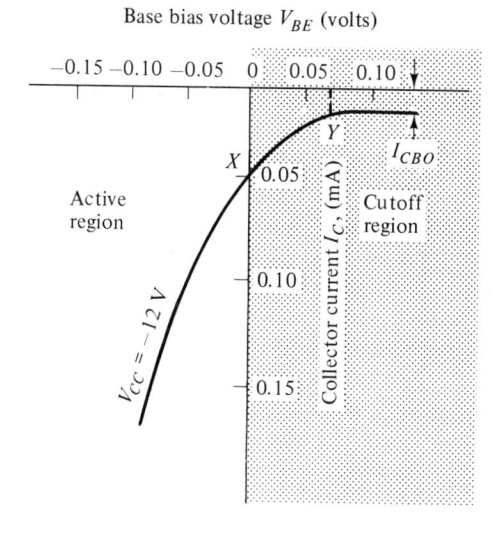

(b) Effect of base bias voltage on collector current

Figure 8–12 Transistor characteristic curves, switching applications.

208 · Nonsinusoidal Oscillator Circuit Design Principles

(a) Transistor switching circuit

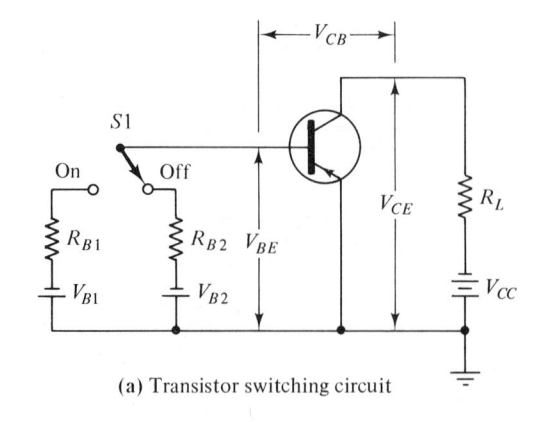

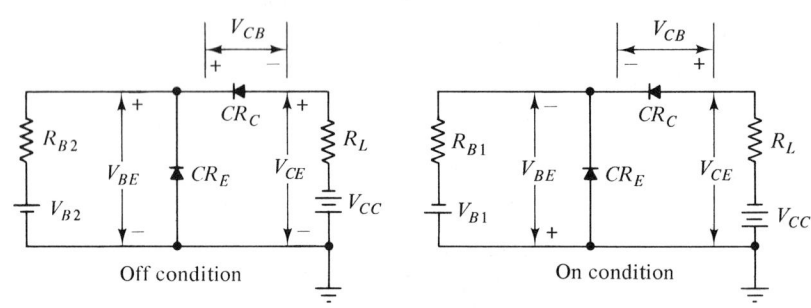

(b) Diode equivalent of transistor switching circuit

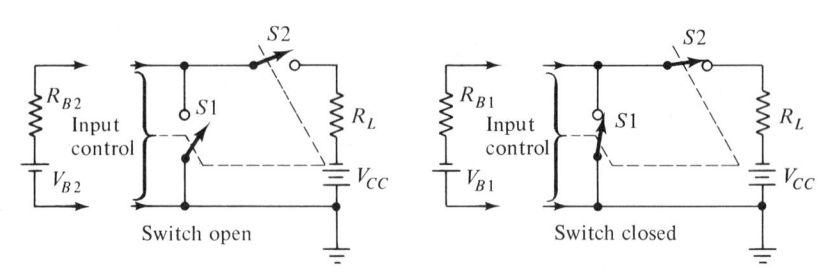

(c) Switch equivalent of transistor switching circuit

Figure 8–13 Transistor switching circuit operation.

current for the CE configuration. Note that the application of a small reverse bias base voltage V_{BE} (approximately 0.075 volt, positive for a PNP transistor) reduces the value of reverse-bias collector current to the value of I_{CBO}.

The collector voltage V_{CE} is indicated by the vertical projection from point Y in Fig. 8–12(a) to the collector-voltage axis. This value is equal to the difference in magnitude between the battery voltage (12 volts in this instance) and the voltage drop produced by reverse-bias collector-current flow through R_L, as in Fig. 8–13(a). Normal quiescent conditions for a transistor switch in this region require that both the emitter-base junction and the collector-base junction must be reverse-biased. With switch S1 in Fig. 8–13(a) in its *off* position, the emitter-base junction is reverse-biased by V_{B2} through R_{B2}. This is comparable to the application of a positive unit step voltage. The collector-base junction is reverse-biased by V_{CC} through R_L; thus, the transistor is in its *off* (cutoff) state.

In the diode equivalent circuit of the transistor, shown in Fig. 8–13(b), diodes CR_E and CR_C represent the emitter-base and collector-base junctions, respectively. Diode CR_E is reverse-biased by voltage V_{BE}; diode CR_C is reverse-biased by voltage V_{CB}. Ideally, there is no current flow through R_L and collector-emitter voltage V_{CE} (output voltage) equals V_{CC}. This circuit, as a switch, is *open*, as depicted in Fig. 8–13(c). The initially applied bias causes switches S_1 and S_2 (which represent the diode equivalents) to open the output circuit.

Next, consider the *active region*. The active linear region in Fig. 8–12(a) is the only region that provides normal amplifier gain. In the linear region, the collector-base junction is reverse-biased and the emitter-base junction is forward-biased. *Transient response of the output signal is essentially determined by transistor characteristics in this region.* In switching circuits, this region represents the transition region. Operation of switch A1 in Fig. 8–13(a) to its *on* position is comparable to the application of a negative unit step voltage. Forward bias is established by V_{B1}, via R_{B1}, on the emitter-base junction, and the base current I_B and collector current I_C become transitory in nature, and move from point X in Fig. 8–12(a) on the load line to point Y. Here, collector current reaches saturation. The signal passes through this region rapidly. *In switching circuit operation, this region is of importance only for design considerations.*

Consider next the *saturation region*. In the saturation region depicted in Fig. 8–12(a), an increase in base current *does not* cause an appreciable increase in collector current I_C. At point Y on the load line, the transistor is in its saturation region. Collector current I_C (measured by a horizontal projection from point Y) is at a maximum, and collector voltage V_{CE} (measured by a vertical projection from point Y) is at a minimum. This value of collector voltage is referred to as the saturation voltage (V_{SAT}), and is an important char-

210 · Nonsinusoidal Oscillator Circuit Design Principles

acteristic of the transistor. *Deep saturation is generally avoided by circuit designers because of its effect on the transient response of the transistor.* The saturation region is also referred to as the *bottomed* region.

When collector current I_C reaches its limited value [battery voltage V_{CC} in Fig. 8–13(a) divided by the load resistance R_L], the transistor saturates and is in its *on,* or conducting, state. In the diode equivalent circuit of the saturated transistor, depicted in Fig. 8–13(b), diodes CR_E and CR_C are forward-biased. Diode CR_E is forward-biased by input voltage V_{BE}. Saturation (output) voltage V_{CE} drops to a smaller negative value than input voltage V_{BE}; the difference in potential of these voltages causes the forward bias V_{CB} on diodes CR_C. The equivalent switch circuit depicted in Fig. 8–13(c) is *closed.* Switches S_1 and S_2 close the circuit for battery voltage V_{CC} through load resistor R_L. Note that the circuit may also be switched from the *on* state to the *off* state in a similar manner. Small input voltage or current pulses may be used to control large output voltage or current pulses. For example, an input voltage swing of approximately 0.2 volt causes an output voltage swing of approximately 11.1 volts for a particular transistor. Correspondingly, an input current swing of 180 µA results in an output current swing of approximately 2800 µA. An average current gain α_{FE} of approximately 15.5 is obtained.

8–8 UJT RELAXATION OSCILLATOR DESIGN

A unijunction transistor has certain advantages in relaxation oscillator operation, as follows: a stable triggering voltage that is a fixed fraction of the applied interbase voltage; a very low value of firing current; a negative-resistance characteristic that is highly uniform from one unit to another, with little temperature dependency; long life expectancy; high pulse current capability. A basic UJT relaxation-oscillator arrangement is shown in Fig. 8–14. To start, the emitter is reverse-biased and is effectively an open circuit. In turn, capacitor C_T charges via R_T to form the ramp of a sawtooth waveform; this is the initial portion of an exponential curve. As the voltage across the charging capacitor reaches the peak-voltage point, the emitter becomes forward-biased and the emitter-to-base resistance suddenly drops to a very low value. In turn, the capacitor discharges quickly, thereby developing the retrace interval. The operating cycle then repeats.

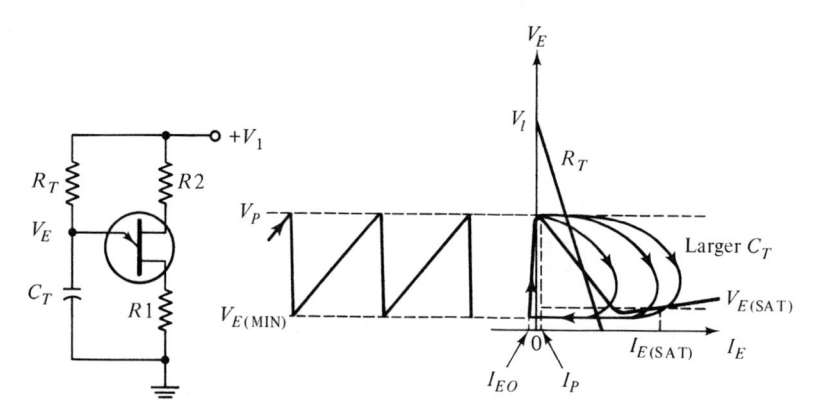

Figure 8-14 Typical UJT sawtooth oscillator configuration, with operating waveforms.

There are several circuit-action factors that the circuit designer should keep in mind: First, the load line for R_T must intersect the emitter characteristic curve to the right of its peak point (this ensures the availability of sufficient trigger current). Note that the *worst-case conditions* involve the upper tolerance of peak current of the emitter characteristic, the lower tolerance on the supply voltage, and the maximum value of peak current at minimum operating temperature. Second, the load line for R_T must intersect the emitter characteristic to the left of its valley point; otherwise, the UJT may "lock up" in its conducting mode. Third, there is a practical limitation on the value of C_T. As the value of C_T is reduced, the amplitude of the output waveform decreases. In turn, the frequency stability of the circuit is adversely affected.

CHAPTER 9

AF and RF Power Amplifier Fundamentals

9-1 GENERAL CONSIDERATIONS

There is no sharp dividing line between small-signal amplifiers and power amplifiers. Driver amplifiers occupy an intermediate position, for example, between a preamplifier and an output power amplifier, or between an oscillator and an output power amplifier, as depicted in Fig. 9-1. The chief distinction between an AF power amplifier and an RF power amplifier is that the latter is tuned. A secondary distinction is that the former operates typically in class AB, whereas the latter usually operates in class C. Push-pull configurations are generally utilized. However, single-ended and push-pull class A power amplifiers are employed in equipments in which large power output is not required and high power efficiency is not a dominating factor. Class B push-pull power amplifiers are used mainly in equipments requiring high power output and high power efficiency.

9-2 THERMAL RUNAWAY HAZARDS

Power amplifiers almost invariably require transistor operation at power levels that are near the runaway condition. This hazard is typically aggravated by use of biasing networks that have marginal stability and which may be required for operating efficiency. Because thermal runaway in power stages is likely to cause catastrophic de-

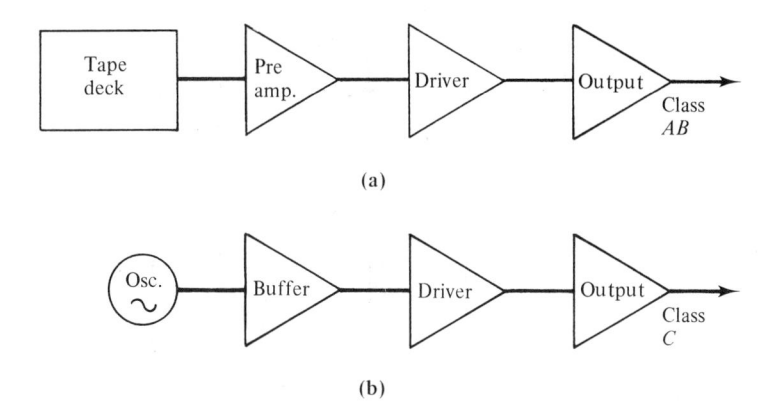

Figure 9–1 Examples of power-amplifier arrangements: **(a)** Tape deck, preamplifier, driver amplifier, and output amplifier; **(b)** Oscillator, buffer amplifier, driver amplifier, and output amplifier.

struction of the transistors, *the circuit designer must be guided by worst-case design principles to eliminate, or at least to minimize, the possibility of thermal runaway.* Worst-case conditions involve the onset of indefinite increase in the value of h_{fe}, zero base-emitter voltage, minimal load impedance, and I_{CO} at a maximum value. In a class B power amplifier, maximum transistor power dissipation occurs at the time when the signal power output is 40 percent of its maximum value; the power dissipated by each transistor is then 20 percent of the maximum power output. On the other hand, in a class A amplifier, maximum transistor power dissipation occurs in the absence of an applied signal.

9–3 COMPLEMENTARY SYMMETRY AMPLIFIER OPERATION

There has been a marked trend to the use of complementary symmetry design in audio power amplifiers. Advantages of this configuration are (1) it reduces circuit complexity; (2) it eliminates a separate phase-inverter stage; (3) it provides outstanding DC stability; (4) it provides extended frequency response, owing to reduced common-mode conduction. A basic configuration is shown in Fig. 9–2. Transistor Q1 is a PNP type, and Q2 is an NPN type in a common-collector configuration. A negative-going input signal forward-biases Q1 and drives it into conduction. Conversely, a positive-going input

214 · AF and RF Power Amplifier Fundamentals

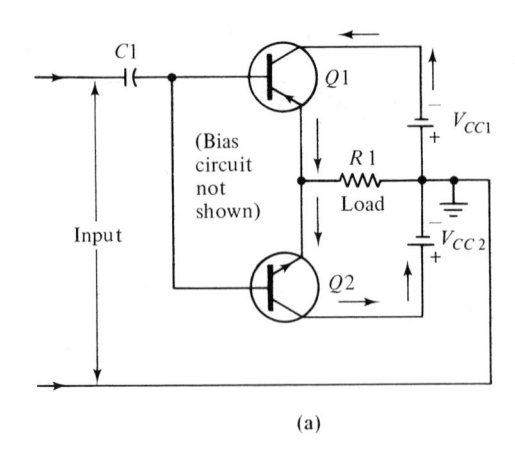

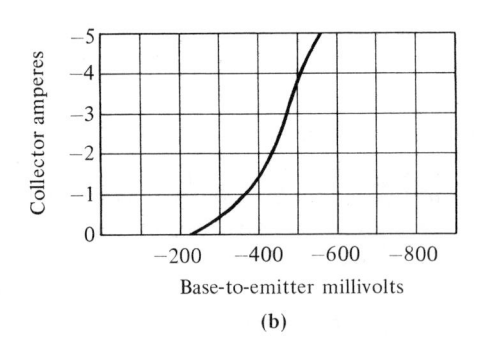

Figure 9-2 A zero-bias complementary-symmetry configuration: **(a)** Skeleton circuit diagram; **(b)** Typical power-transistor transfer characteristic.

signal forward-biases Q2 and drives it into conduction. A signal that forward-biases one transistor will reverse-bias the other transistor. Accordingly, only one transistor conducts at any given time. Resultant action in the output circuit can be understood by consideration of the circuit depicted in Fig. 9-3. This is a simplified version of the output arrangement. The internal emitter-collector circuit of Q1 is represented by variable resistor R1, and that of Q2 is represented by variable resistor R2.

With no input signal, and in class B operation (zero emitter-base bias), the arms of the variable resistors are regarded as in their *off* positions. No current flows through the transistors nor through the load resistor R_L. As the incoming signal swings positive, Q2 con-

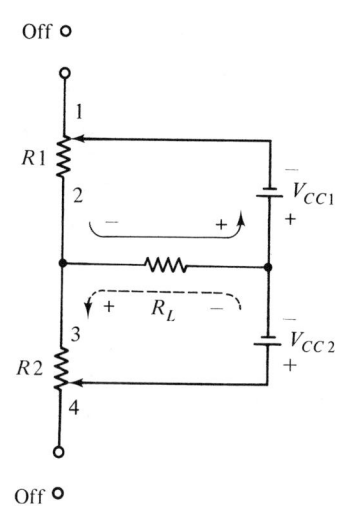

Figure 9-3 Simplified version of a complementary symmetry output circuit.

ducts and Q1 remains nonconducting. Variable resistor R1 remains in its *off* position. The arm of R2 moves toward point 3, and current flows through the series circuit consisting of battery V_{CC2}, resistor R_L, and variable resistor R2. The amount of current flow depends upon the magnitude of the incoming signal, with the arm moving toward point 3 for increased forward bias, and toward point 4 for decreased forward bias. The current flows in the direction of the dashed arrow, producing a voltage with the indicated polarity.

When the input signal goes negative, Q1 conducts and Q2 becomes nonconducting. The same action is repeated with respect to variable resistor R1. Current flows through V_{CC1}, variable resistor R1, and load resistor R_L in the direction shown by the solid-line arrow, and produces a voltage across R_L of the indicated polarity. *It follows that the two transistors must have closely matched characteristics, or the output waveform will be distorted.* For example, if the beta value of one transistor is 20 percent greater than the beta value of the other transistor, the output will contain serious even-harmonic distortion. Note also that a zero-bias class B stage tends to develop crossover distortion as a consequence of nonlinear transistor operation at low signal levels.

For class A operation, the configuration depicted in Fig. 9-2 is elaborated by application of sufficient forward bias to the two transistors that collector current is never cut off at any time. In the sim-

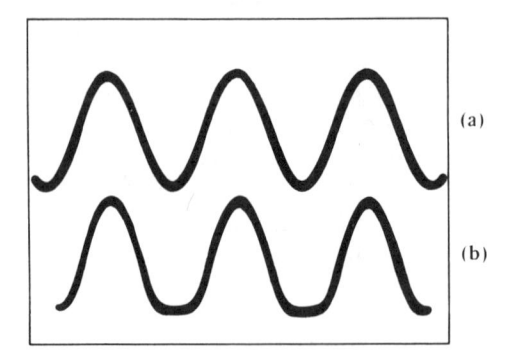

Figure 9–4 Complementary-symmetry amplifier output waveforms: **(a)** Balanced output; **(b)** Unbalanced output.

plified circuit of Fig. 9–3, the variable resistors will not be in the *off* position at any time. Bias current in the output circuit flows out of the negative terminal of V_{CC2} into the positive terminal of V_{CC1} via variable resistor R1, variable resistor R2, and into the positive terminal of V_{CC2}. Under these conditions, the output circuit can be considered a balanced bridge; its arms consist of R1, R2, V_{CC1}, and V_{CC2}. When the input signal goes positive, Q2 conducts more and Q1 conducts less. In the simplified circuit, the arm of R1 moves toward point 1, and the arm of R2 moves toward point 3. In turn, the bridge becomes unbalanced and electrons flow through R_L in the direction of the dashed-line arrow, producing a voltage drop of the indicated polarity.

When the input signal goes negative, Q1 conducts more and Q2 conducts less. In the simplified circuit, the arm of R1 moves toward point 2, and the arm of R2 moves toward point 4. The bridge becomes oppositely unbalanced, and electrons flow through R_L in the direction of the solid-line arrow, producing a voltage drop of the indicated polarity. In either class A or class B operation, DC current does not flow through the load resistor. Thus, if the voice coil of a speaker is connected directly in place of R_L, the voice coil will not be offset by DC current flow. In the event that the circuit is not balanced with respect to DC current, distortion will be introduced into the output signal (see Fig. 9–4). Thus, Q1 and Q2 should have closely matched characteristics. The complementary symmetry configuration is one of the basic output transformerless (OTL) arrangements.

Most complementary symmetry stages are operated in class

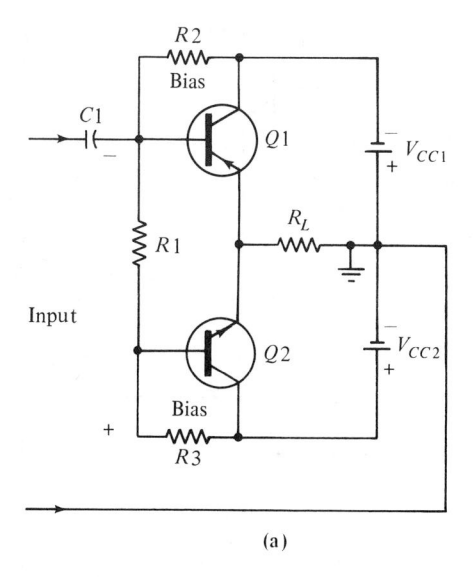

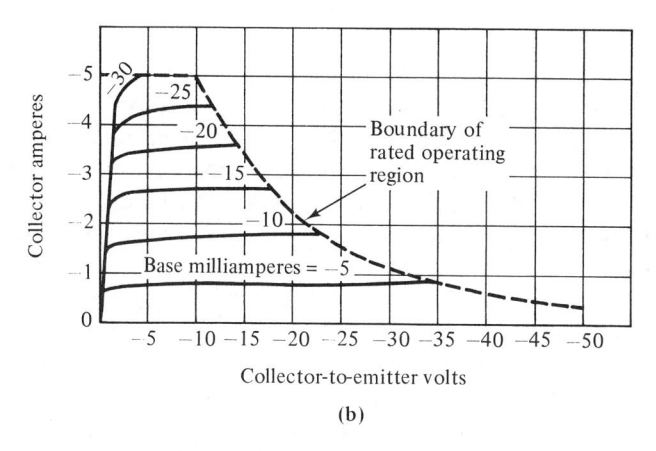

Figure 9–5 Forward-biased CE complementary symmetry configuration: **(a)** Schematic diagram; **(b)** Maximum power dissipation rating for a typical power transistor.

AB. Sufficient forward bias is employed to minimize crossover distortion. Note that too much forward bias in a class-AB amplifier introduces another type of distortion termed "stretching" distortion. Thus, the optimum bias value is comparatively critical. Observe the circuit depicted in Fig. 9–5. Here, a small forward bias has been ap-

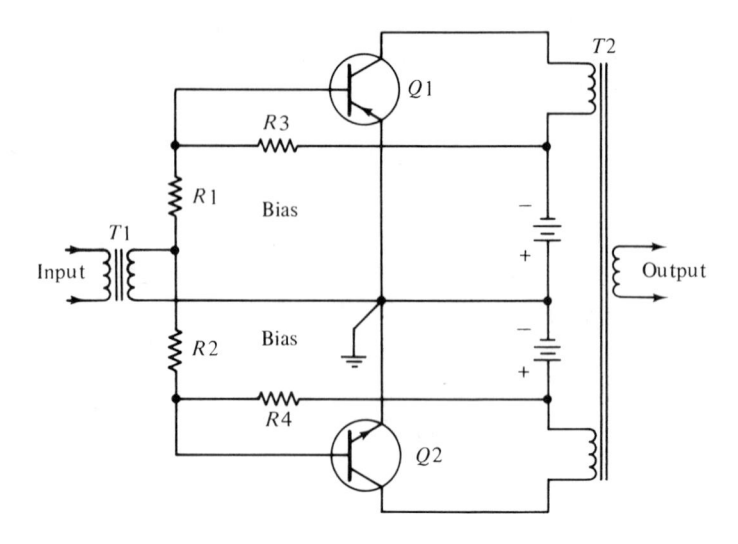

Figure 9-6 Complementary symmetry circuit with transformer-coupled input and output.

plied. This is a common-collector configuration, similar to the arrangement depicted in Fig. 9-2. However, a voltage divider (bias circuit) comprising R1, R2, and R3 connected in series across V_{CC1} and V_{CC2} has been added in the configuration of Fig. 9-5. Forward bias for both transistors is provided by the voltage drop across R1.

In one of the bias loops, R1 is in series with the base-emitter junctions of the two transistors. If the DC resistances of the two junctions are considered to be equal to each other, the voltage drop across each junction is half of that developed across R1. Normally, the value of R1 is very small, so that its unbalancing effect on the input signal to Q2 is negligible. *It follows that optimum class AB operation can be obtained only if the characteristics of Q1 and Q2 are reasonably well matched.* Substantial mismatch can result in a combination of crossover distortion and stretching distortion in the output signal. It can also result in excessive power dissipation in one transistor, and less than normal dissipation in the other transistor.

There has been a marked trend away from transformer coupling. However, it is instructive to observe the common-emitter complementary symmetry circuit with transformer input and transformer output coupling exemplified in Fig. 9-6. This arrangement operates in class AB to minimize crossover distortion. The voltage divider consisting of resistors R3 and R1 establishes the forward

Complementary Symmetry Amplifier Operation · 219

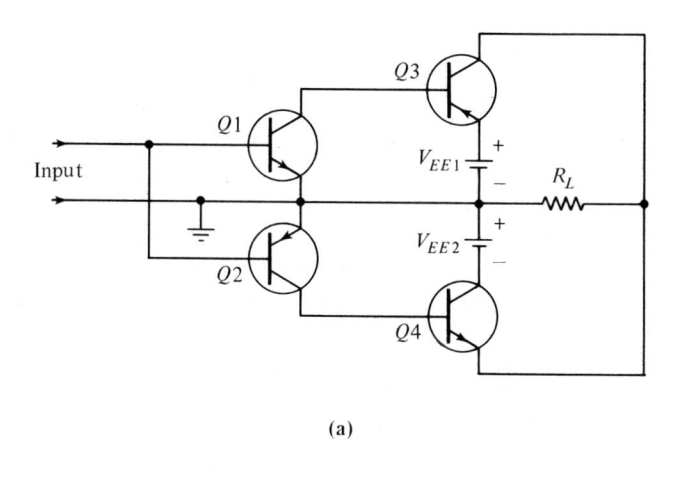

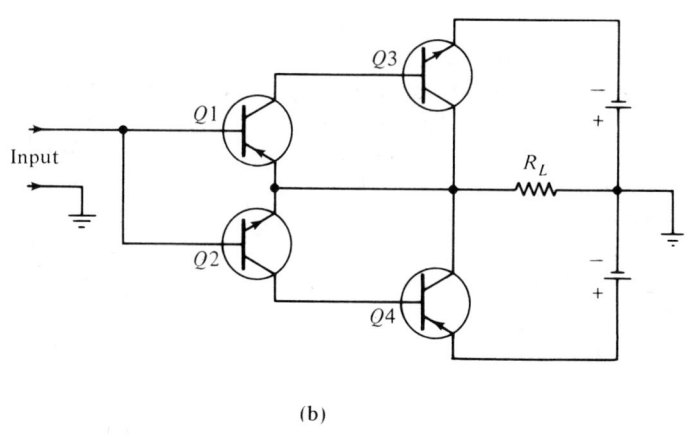

Figure 9–7 Examples of direct-coupled complementary symmetry stages.

bias on Q1. Transistor Q2 is forward-biased by the voltage divider consisting of resistors R4 and R2. Note that a tapped-secondary transformer is not required at the input. However, a split-primary output transformer (T2) must be utilized. Since NPN and PNP transistors are employed, opposite polarity voltages are required for the collectors.

Cascaded CE Complementary Symmetry Arrangements Cascaded common-emitter complementary symmetry stages are in common use. Refer to Fig. 9–7. One common-emitter complementary symmetry stage (Q3 and Q4) is directly driven by another common-emitter complementary symmetry stage (Q1 and Q2). The input to Q1 and Q2 is from a single-ended stage (not shown). When the input signal goes positive, Q1 conducts and Q2 remains nonconducting. Because of 180-deg phase reversal in a *CE* configuration, the collector of Q1 goes negative, which causes Q3 to conduct and Q3's collector goes positive. When the input signal goes negative, Q2 conducts and its collector goes positive. This causes Q4 to conduct and its collector goes negative. Transistors Q1 and Q3 are nonconducting during this period. Battery V_{EE1} supplies the required biasing voltages for transistors Q1 and Q3. Battery V_{EE2} supplies the required biasing voltages for transistors Q2 and Q4.

The emitter-base junction of Q3 is in series with the collector-emitter circuit of Q1 and V_{EE1}. As a result, the emitter of Q3 is positive with respect to its base (forward bias), and the collector of Q1 is positive with respect to its emitter, as is required for electron flow through Q1. A similar function exists for Q2 and Q4 with battery V_{EE2}. A circuit similar in operation, but with a different arrangement is exemplified in Fig. 9-7(b). Transistors Q1 and Q2 are connected to provide a high input resistance. *Signal voltage developed across R_L provides a negative feedback in the input circuit of Q1 and Q2, to develop a high input resistance.* Transistors Q3 and Q4 are connected in a *CE* configuration to match the output resistance of Q1 and Q2. Just as in the case of amplifiers with a single transistor, use of the common-emitter configuration in *the cascaded complementary symmetry configuration provides high power gain to a low-impedance load.*

Compound-connected Complementary Symmetry Arrangement
As noted previously, the current gain, voltage gain, and power gain of a transistor are directly proportional to the short-circuit forward-current amplification factor. This factor is the ratio of the output current to the input current, measured with the output short-circuited. To obtain maximum gain in a given amplifier, it is necessary to use a transistor that has a high-valued short-circuit forward-current amplification factor. Most transistors have an α_{fb} ranging from 0.940 to 0.985, with an average value of 0.96. Regardless of the value of α_{fb}, this factor will decrease as the emitter current increases. Variation of collector current with emitter current for a single transistor is exemplified in Fig. 9-8.

The ratio of collector current to emitter current remains constant (straight line interval) to approximately 400 mA; then the collector current increases less rapidly (curved interval) as the emitter

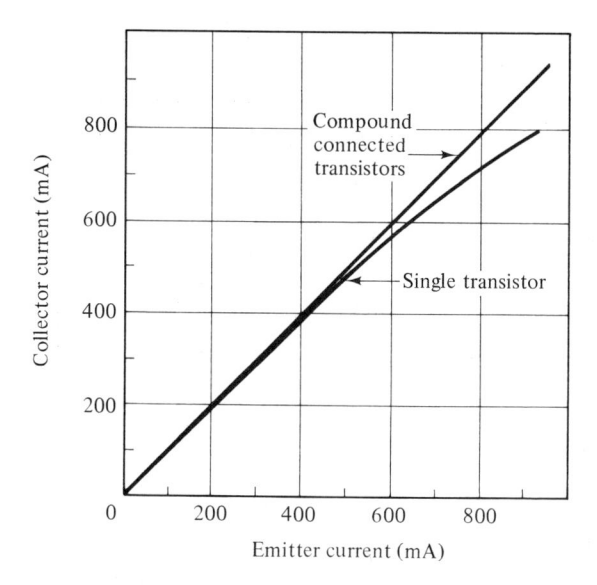

Figure 9-8 Variation of current gain for a single transistor, and for a pair of compound connected transistors.

current increases. This indicates a reduction in the current amplification factor at high values of emitter current. In power amplifiers that draw heavy emitter current, this variation is aggravated. If two transistors are compound-connected, as depicted in Fig. 9-9, the dropoff of collector current (or reduction of current gain) at high emitter currents can be made negligible. The dashed lines in Fig. 9-9 enclose two compound-connected transistors. Note that the base of one transistor is connected to the emitter of the other transistor, and the two connectors are connected together. In this circuit, the transistors are operated in the CB mode.

It can be shown that the current gain for the compound-connected transistors is greater than for a single transistor. For example, suppose that a pair of transistors each has a current-gain value of 0.95; when they are connected in compound, their combined current-gain value becomes 0.9975. *This increase is equivalent to a beta value increase from 19 to 399.* In other words, the effectiveness of the compound connection is better indicated by considering the short-circuit forward-current amplification factor in the CE configuration. Compound-connected transistors in a circuit of any configuration can be employed and considered as a single unit with a high current-amplification factor.

222 · AF and RF Power Amplifier Fundamentals

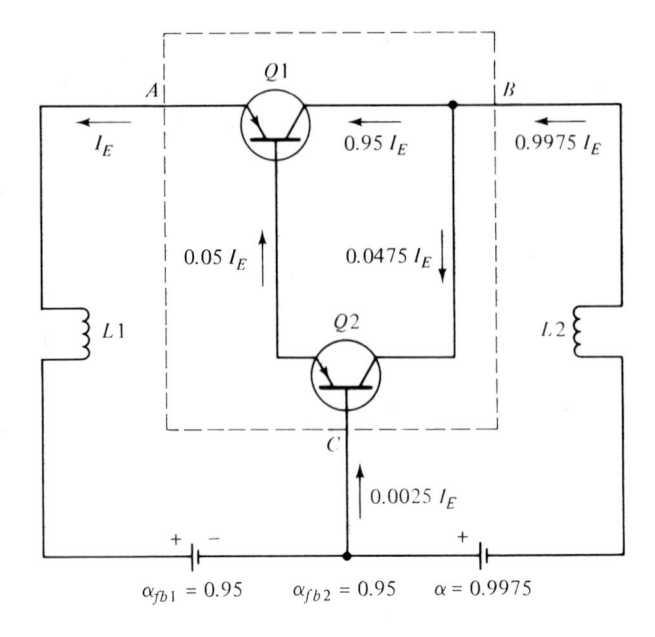

Figure 9–9 Flow diagram for compound-connected transistors.

Note that the foregoing discussion stipulated transistors that had equal alpha (or beta). However, this equality should not be regarded as mandatory. That is, the advantages of the compound connection are not restricted to transistors that have equal-valued current-amplification factors. Compound-connected transistors can be utilized in single-ended amplifiers, in conventional push-pull amplifiers, or in complementary-symmetry amplifiers. An example of the latter arrangement is shown in Fig. 9–10. This circuit is similar to the complementary-symmetry configuration depicted in Fig. 9–5. The junction of the correspondingly referenced parts is the same. However, Q1 is replaced by the compound connection of Q1A and Q1B in Fig. 9–10. Transistor Q2 is replaced by the compound connection of Q2A and Q2B. *This configuration is also called a Darlington-pair complementary-symmetry arrangement, and is in extensive use.*

Complementary-symmetry Bridge Configuration A basic bridge circuit arrangement is illustrated in Fig. 9–11. The arms of the bridge, W, X, Y, and Z, can be circuit elements such as resistors, in-

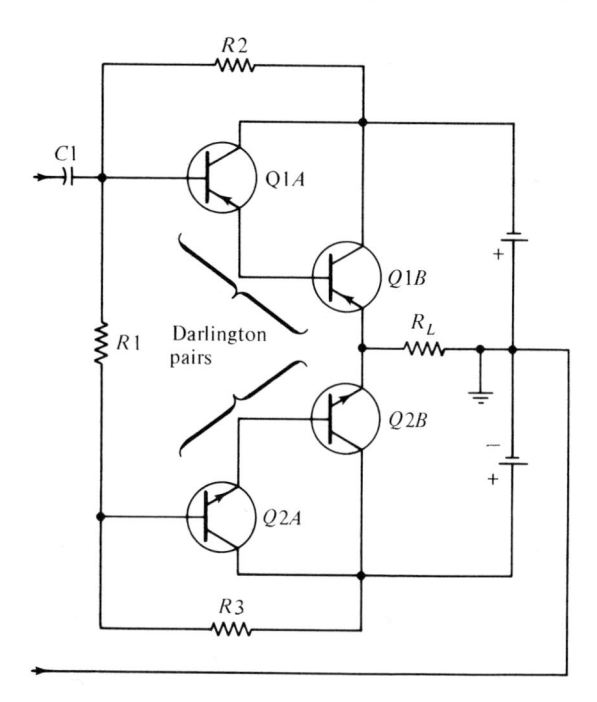

Figure 9-10 Complementary-symmetry compound-connected (Darlington) output-amplifier arrangement.

ductors, capacitors, transistors, or batteries. Bridge balance occurs when, regardless of the voltage applied at points 1 and 2, the voltage drops across the bridge arms are such that zero voltage is developed across points 3 and 4; that is, no current flows through R_L. The arms of the basic bridge can be replaced by two transistors (Q1 and Q2) and two batteries (V_{CC_1} and V_{CC_2}), as shown in Fig. 9-11(b). If the two transistors are biased to draw equal emitter currents (or no emitter-collector current) under quiescent conditions (no input signal), the bridge is balanced and current flows through R_L. However, if an input signal causes either transistor to conduct more than the other, the bridge becomes unbalanced and current flows through R_L. Sine-wave input signals fed to points *A-B* and *C-B* 180 deg out of phase will produce an amplified sine wave signal across R_L. If the transistors are operated in class A or class B, no DC current will flow through R_L.

The load resistor can be replaced by a speaker voice coil. As noted previously, DC current flow through a speaker voice coil

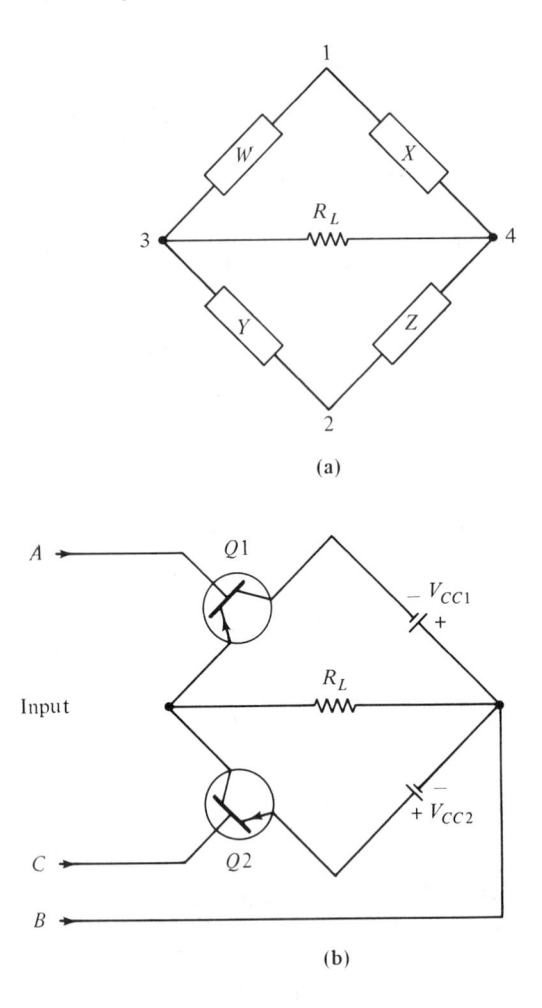

Figure 9-11 A basic bridge arrangement, and a bridge with two transistors and two batteries.

causes offset and resulting distortion in the output signal. Note that if a conventional push-pull amplifier were employed, the speaker voice coil would have to be center-tapped and only half of the coil would be used for each half-cycle of the input signal. This entails reduced efficiency in converting electrical energy into sound energy. Observe the bridge configuration illustrated in Fig. 9-12, in which the bridge arms comprise transistors. Transistors Q1 and Q2 are PNP types, whereas Q2 and Q4 are NPN types. The advantage of

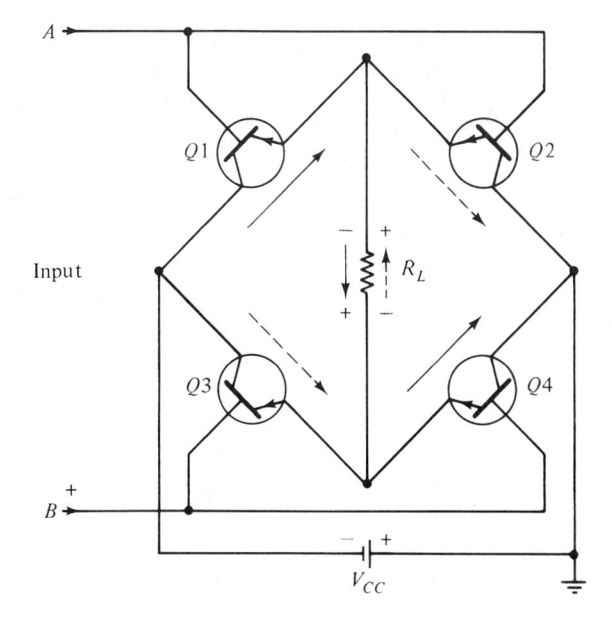

Figure 9–12 Complementary-symmetry bridge arrangement.

this arrangement is that no part of the input circuit or the output circuit need be placed at ground potential. Moreover, this configuration can be driven by a single-ended driver; in other words, two input signals 180 deg out of phase are not required.

Assume that under quiescent conditions all of the transistors are zero-biased and draw no current (class B operation). If the transistors draw no current, there is no completed circuit across V_{CC} and no DC current flows through R_L. Next, if an input signal causes point A to become negative with respect to point B, Q1 and Q4 become forward-biased, and Q3 and Q2 become reverse-biased. Transistors Q1 and Q4 conduct. Electrons flow out of the negative battery terminal through Q1 collector-to-emitter, through R_L, through Q4 emitter-to-collector, and back to the positive battery terminal. This path is indicated by the solid-line arrows. A voltage is developed across R_L with the indicated polarity.

If an input signal causes point A to become positive with respect to point B, Q3 and Q2 conduct, whereas Q1 and Q4 are nonconducting. The electron current path is indicated by the dashed-line arrows. A voltage is developed across R_L in the polarity indicated. Assume that under quiescent conditions the transistors are biased to draw equal currents in class A operation. (This bias circuit is not shown.) Under this condition, electrons emerge from the negative battery terminal. One-half of the electron current flows through Q1 collector-to-emitter and through Q2 emitter-to-collector, and thence into the positive battery terminal. The other half of the electron current flows through Q3 collector-to-emitter and through Q4 emitter-to-collector and thence back into the positive battery terminal. Thus, the bridge is balanced and there is no current flow through R_L.

Next, assume that an input signal causes point A to become more negative with respect to point B. Transistors Q1 and Q2 become more forward-biased and draw more collector current. Transistors Q2 and Q3 become less forward-biased and draw less collector current. Thus, the bridge becomes unbalanced, and the difference in current between Q1 and Q2 flows through R_L in the direction of the solid-line arrow and through Q4 into the positive battery terminal. A voltage is developed accordingly across R_L with the indicated polarity. If the input signal causes point A to become positive with respect to point B, Q2 and Q3 become more forward-biased and draw more collector current. Transistors Q1 and Q4 become less forward-biased and draw less collector current. The difference in current between Q3 and Q4 flows through R_L in the direction of the dashed-line arrow and develops a voltage with the indicated polarity.

Summary of Amplifier Design Considerations

1. The noise factor of an amplifier is equal to the quotient of the signal-to-noise ratio at the output of the amplifier divided by the signal-to-noise ratio at the input of the amplifier.
2. As the collector voltage of a transistor is increased, its noise factor also increases.
3. A CE amplifier with degeneration develops a comparatively high input impedance.
4. Zero-biased class B push-pull amplifiers produce crossover distortion.
5. Crossover distortion can be minimized or eliminated by operating a push-pull stage in class AB.
6. Stretching distortion develops if excessive forward bias is utilized in a class AB amplifier.
7. Complementary symmetry push-pull amplifiers have numerous advantages and are widely used. Transistors must have closely matched characteristics to avoid distortion.

8. Compound-connected or Darlington-connected transistors provide a high current amplification factor.
9. Darlington-connected transistors are frequently used in complementary-symmetry circuits for high gain and power output with minimum circuit complexity.

9–4 RF POWER AMPLIFIERS

RF power-amplifier design differs from AF design primarily in the employment of tuned coupling circuits. This feature is essential to avoid prohibitive loss of RF energy owing to bypassing action of junction capacitances and stray circuit capacitances. In other words, the inherent circuit capacitances are utilized in tuning inductors or transformers to the amplifier operating frequency. In turn, the otherwise low reactance of a junction capacitance, for example, is converted into a comparatively high impedance (at the operating frequency) by parallel-resonant circuit action. Inasmuch as power transfer is a dominant design consideration, impedance matching is maintained both at the input and at the output of an RF power amplifier. Waveform distortion is of little concern, because harmonics are largely rejected by the tuned circuitry.

An RF power amplifier may be operated in either class B or in class C. A class B amplifier may be designed to provide an RF output signal that is proportional in amplitude to the input signal level; it is often called a *linear* RF amplifier. Most RF power amplifiers are operated in class C, in order to realize maximum efficiency. A typical class C RF power amplifier has an efficiency of 65 to 70 percent. This is the ratio of the RF power output to the DC power input of the transistor(s). For example, if the bogie RF power-output value is 50 W, the DC power input will be approximately 70 W. Thus, if a 28-V collector supply is utilized, the collector current will be 2.7 A. This example is simplified to the extent that the RF driving power to the transistor is ignored. Note that if the collector circuit is tuned to a multiple of the driver input circuit, the stage operates as a *frequency multiplier*. A doubler has a typical efficiency of 40 percent; a tripler operates at approximately 28 percent efficiency; a quadrupler has an efficiency of about 21 percent, and a quintupler operates at approximately 18 percent efficiency.

A basic RF power-amplifier configuration is shown in Fig. 9–13. A relatively high driver signal is fed into the input of the power-amplifier stage. This is a typical series-fed arrangement. *Use*

228 · AF and RF Power Amplifier Fundamentals

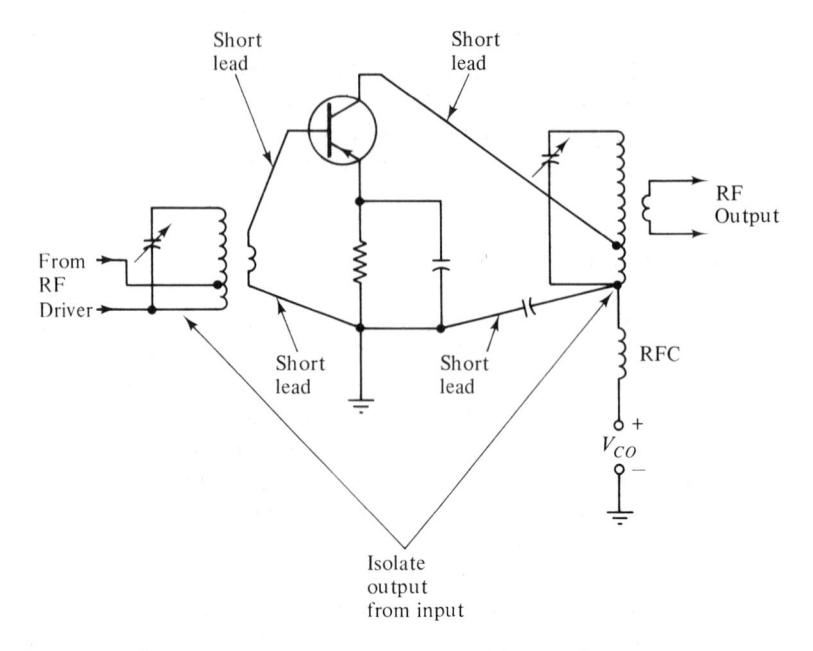

Figure 9–13 Basic RF power-amplifier configuration.

of shunt feed may result in parasitic oscillations that are difficult to eliminate. If parasitic oscillation occurs in the configuration of Fig. 9–13, a selected RF choke should be used, or a ferrite bead may be placed on the pigtail lead of the choke. The emitter swamping resistor may have a value of 1 ohm. Signal leads must be kept as short as possible, because even a straight piece of wire has appreciable inductance at the higher radio frequencies. This is essentially a class B circuit that will operate in class C when a substantial drive signal is applied. Reverse bias developed across the emitter swamping resistor charges the emitter bypass capacitor. This capacitor has a typical value of 0.1 μF.

Sometimes parasitic oscillations can be controlled only with a neutralizing capacitor, as exemplified in Fig. 9–14. Resistor R_B assists in avoiding parasitic oscillation involving the base RF choke. A ferrite bead on the pigtail lead on the choke may also be helpful. In various situations, the circuit designer must use selected RF chokes to avoid parasitic oscillation. Note that this is a class B amplifier insofar as a small signal input is concerned. However, with substantial drive applied, base-emitter rectifier action charges C_C and reverse-biases the transistor for class C operation.

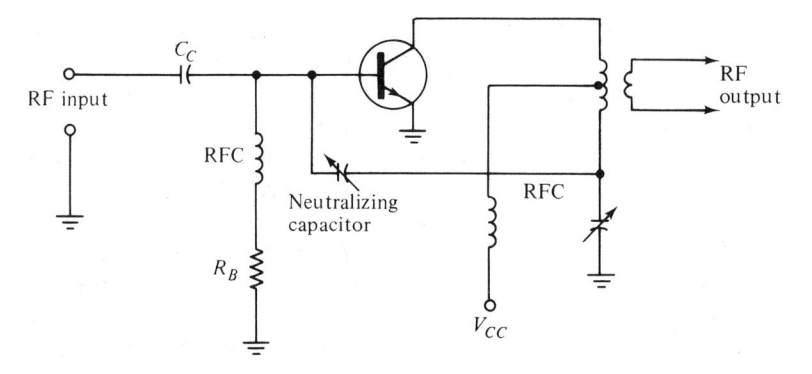

Figure 9-14 Neutralizing capacitor in an RF power-amplifier arrangement.

In some applications, extremely good input/output linearity may be required. In such a case, the stage would be operated in class A. Although the power gain is considerably greater than in the class B or class C mode, the operating efficiency is only 25 percent, approximately. In other applications, such as in single-sideband (SSB) transmitters, class B push-pull operation is commonly used for good linearity and reasonable efficiency. In both class A and class B modes of operation, the designer must exert greater care to prevent thermal runaway than in class C operation. Note that when class C bias is obtained with an emitter swamping resistor, as exemplified in Fig. 9-13, the emitter bypass capacitor may present a problem in high-frequency operation. Circuit designers typically use several capacitors connected in parallel to minimize the lead inductance. Alternatively, the lead inductance may be critically adjusted for series resonance with the capacitor.

Good practices in RF circuit design include adequate bypassing of power-supply leads, preferably with two paralleled capacitors. One of the capacitors should have a high value in order to offer a low reactance at low (potentially parasitic) frequencies; the transistor ordinarily has relatively high gain at low frequencies. The other capacitor should have a comparatively small value with a low RF reactance. This small capacitor is required because a large capacitor has excessive series impedance at high radio frequencies. Sintered-electrode tantalum capacitors are suitable for RF bypassing up to approximately 75 MHz. These are good general-purpose RF bypass capacitors. However, high-Q ceramic capacitors are preferred for UHF bypassing.

Components can be checked for self-resonances or parasitic impedances on an RF impedance bridge. *When these potential trouble sources are found, their possible adverse effects on performance of the prototype should be carefully evaluated.* In turn, the circuit designer can make a judicious selection of available inductors and capacitors. Note in passing that feedthrough bypass capacitors and mica "postage-stamp" capacitors have superior characteristics in VHF operation. Resistors that carry RF current should have low series inductance and low shunt capacitance. Small composition resistors are usually satisfactory. However, as noted previously, stray shunt capacitance can have a drastic action on the effective value of a composition resistor in the megohm range at radio frequencies. When possible, all of the RF grounds should be connected to the ground plane within a small area.

CHAPTER 10

Debugging Logic Systems

10−1 GENERAL CONSIDERATIONS

Logic systems usually require debugging during the design period. Production rejects also require debugging. The most efficient procedure is to use logic state analyzers for fault localization, as explained in the following discussion provided by courtesy of Hewlett-Packard Co. As a generalization, instrumentation enters into the design cycle as soon as hardware is present, whether it is a partially working breadboard or some combination of simulation/emulation equipment and the designer's own hardware. A considerable variety of aids is available to the microprocessor designer, but when the troubleshooting problems become more real-time in nature, more subtle in their interactive hardware/software idiosyncrasies, a real-time noninteractive measurement tool is invaluable. Illustration is still the best way to make a point. Accordingly, it is instructive to consider some fairly common problems to solve, which are often very troublesome without suitable measurement capability.

Consider time and other parameters versus state flow. In this first example, a case is described in which the designer wished to measure the overall cycle of time of an addressing system in a microprocessor. His case involved address buffering, ROM chip selection logic, and three-state buffering of the data bus. Doing a theoretical calculation of the access time of this addressing scheme yielded at best only hypothetical results. What he desired to know

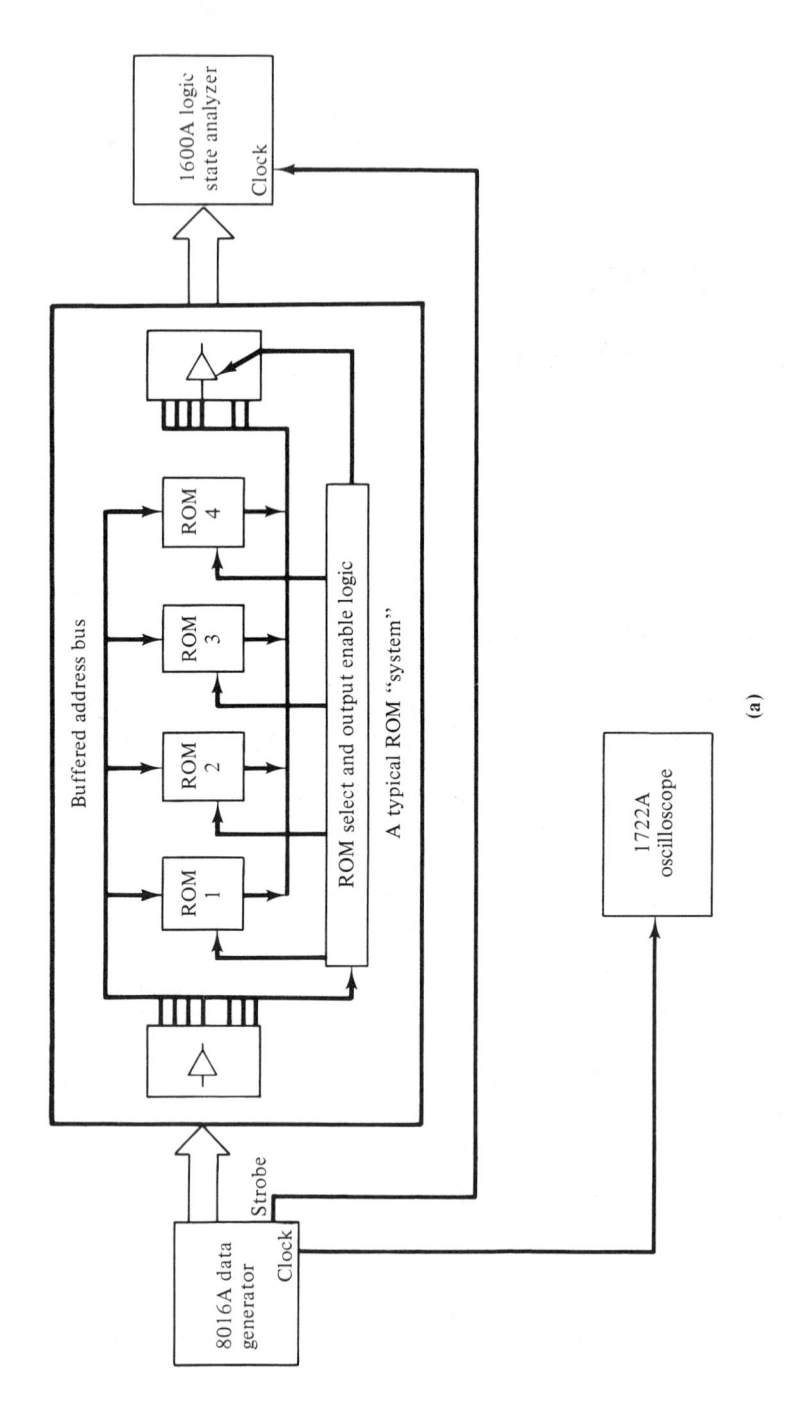

Figure 10–1 Determining actual limits of access time of a complex memory addressing arrangement is best accomplished with stimulus/response testing: **(a)** test arrangement.

(a)

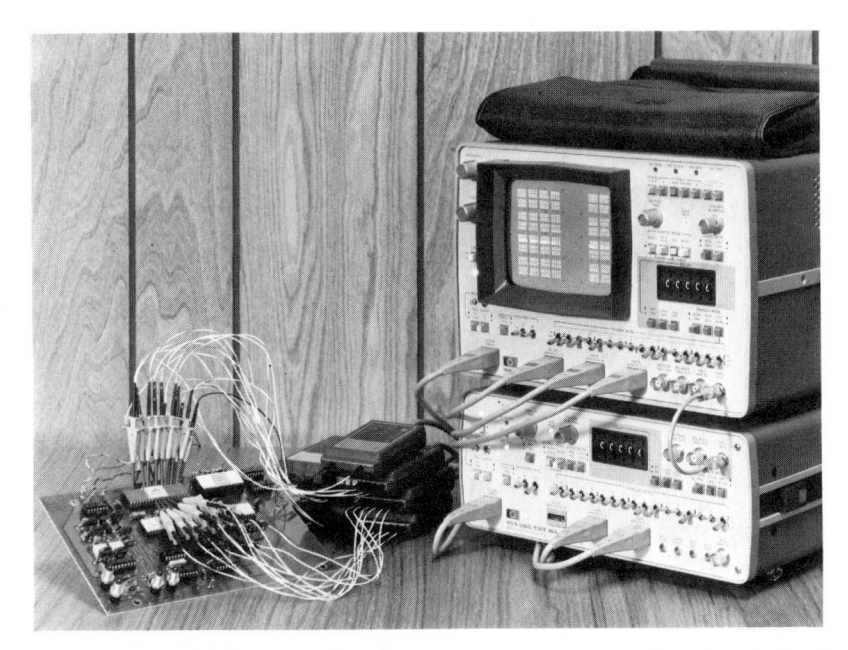

Figure 10-1 *Continued* **(b)** logic state analyzers. *(Courtesy of Hewlett-Packard)*

was just how fast his system really worked. To this end, he instrumented the experiment as shown in Fig. 10-1. In place of the microprocessor, a stimulus was applied that was fully controllable in both frequency and state value so that addresses could be generated at a variable rep rate. In order to measure the response of the system, a data monitor was needed that read the memory output data and indicated when those data were no longer correct. For this he used a Hewlett-Packard logic state analyzer as a response monitor. (This example illustrates stimulus/response testing in a digital world analogous to the stimulus/response testing so commonly used in the analog world.)

For the experiment, he set up the word generator to provide an addressing pattern that he considered to be worst-case. He then ran this addressing pattern at a moderate speed and stored the data (Fig. 10-2) as correctly read by the logic analyzer in the B-memory as a reference. Having stored these data, he used the Exclusive-OR display mode to show if what was currently being read compared exactly to what he stored as a reference. It was then a simple matter to increase the repetition rate of the word generator until the in-

234 · Debugging Logic Systems

```
11 101 010          11 101 010
01 100 011          01 100 011
01 100 100          01 100 100
00 100 111          00 100 111

10 101 011          10 101 011
10 101 100          10 101 100
10 101 111          10 101 111
10 110 001          10 110 001

01 110 011          01 110 011
01 110 101          01 110 101
01 110 100          01 110 100
01 110 111          01 110 111

01 111 001          01 111 001
00 111 000          00 111 000
10 111 011          10 111 011
10 111 110          10 111 110
```
(a)

```
11 101 010          00 000 000
01 100 011          00 000 000
01 100 100          00 000 000
00 100 111          00 000 000

10 101 011          00 000 000
10 101 100          00 000 000
10 101 111          00 000 000
10 110 001          00 000 000

01 110 011          00 000 000
01 110 101          00 000 000
01 110 100          00 000 000
01 110 111          00 000 000

01 111 001          00 000 000
00 111 000          00 000 000
10 111 011          00 000 000
10 111 110          00 000 000
```
(b)

Figure 10–2 (a) An entire data table can be duplicated in the B-memory as a stored reference for comparison with active data; (b) The B-memory can be reduced to an XOR display of the stored data and the active data being strobed into the A-memory; (c) Intensified "l's" in the XOR display indicate that logic differences have been detected between the data in the two memories. *(Courtesy, Hewlett-Packard.)*

```
10  101  010         01  000  000
01  100  011         00  000  000
01  100  100         00  000  000
00  100  110         00  000  001

10  101  011         00  000  000
10  101  100         00  000  000
10  101  111         00  000  000
10  110  001         00  000  000

01  110  011         00  000  000
01  110  101         00  000  000
01  110  100         00  000  000
01  110  110         00  000  001

01  111  001         00  000  000
00  111  000         00  000  000
10  111  011         00  000  000
10  111  110         00  000  000
```

(c)

Figure 10-2 (Cont.)

strument began to detect differences. Then he reduced the frequency just slightly from that point and measured it. The frequency thus read, converted to cycle time, was the actual—not simulated or theoretical—cycle time of the memory addressing system.

The previous experiment is representative of a class of problems dealing with real-time data flow. For instance, there are always such questions as how fast a system can respond to certain stimuli, such as interrupts; or how quickly two keys can be struck in sequence on a terminal or calculator keyboard and still have the data read correctly, and so on. Usually, these operations require several levels of logic and most often require an interaction between software and logic. The way to get the most dependable answers to these kinds of questions is through real-time measurement rather than simulation.

Of course, speed is not the only variable that the designer may want to measure on a parametric basis. He may prefer to look at overall system operation. The map display mode (Fig. 10-3) of a logic state analyzer allows him actually to watch a program being executed. The designer can establish a correct map pattern by photograph or just in his mind's eye, and it becomes very easy to detect when that pattern changes on even a relatively slight basis. In this

236 · Debugging Logic Systems

(a)

(b)

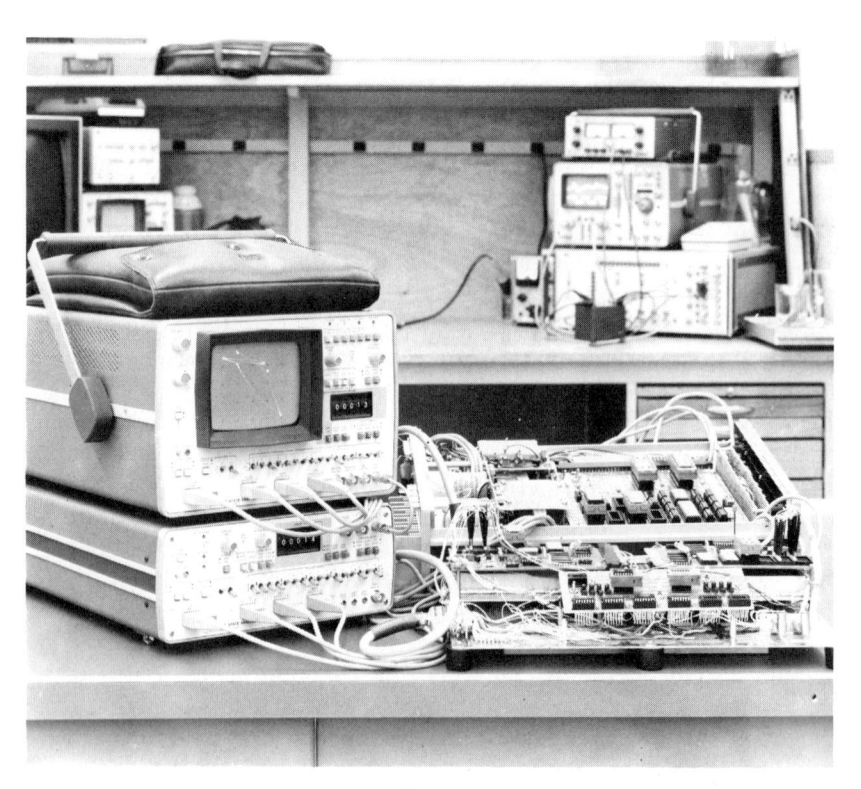

way, the designer can monitor entire operations while he does things such as vary power-supply voltage, inject noise, vary temperature, or create any sort of environment that he wishes. *He can run his prototype or even production units through extensive tests,* to see at what point the operation begins to "blow up." Not only is it instantly recognizable on the map when a machine begins to do something differently, but it is also possible, with the aid of the cursor, to track down the area in which the operation appears to fail first and examine that area in detail with the resultant triggered table display modes.

Sensitivity to noise or ringing on bus lines is of considerable concern to the logic designer. Ringing is a common form of "glitch" that often causes erroneous state flow, but to measure threshold sensitivity in a complex system under all sorts of conditions has been a difficult problem. Quite often the worst problems are present only under unique circumstances. For example, one bit of a 16-bit I/O is read intermittently in error only when nearly all bits in the data switch at once. One method to examine the system's noise sensitivity is to use a Hewlett-Packard 1600S system, consisting of two logic analyzers connected together to form one 32-bit analyzer capable of simultaneously displaying and triggering on two independently operating 16-bit data streams. It can easily capture and display simultaneously address and data up to 32 bits in width, show data being read from the same bus at two different threshold settings, and Exclusive-OR the display to highlight differences, and show dual-clock systems on one screen for real-time analysis of such things as I/O data transfer.

Thereby, the designer can independently read the data at two different threshold settings and actively compare the results. The A-memory inputs should read the data at a variable threshold setting under the control of the operator, while comparing each bit (Exclusive-OR display again) with data being read into the B-memory at the correct threshold setting (Fig. 10–4). The entire program can be scanned by using the pattern triggers and digital delay generators to

◄

Figure 10-3 The map display of a logic state analyzer depicts each 16-bit word as a single dot on screen with a vector drawn between the dots to indicate data sequence. The map allows the designer to view entire operations of his machine, watching for pattern changes that indicate the machine is no longer executing the same operations in the same manner. **(a)** typical map display; **(b)** logic state analyzers. *(Courtesy, Hewlett-Packard.)*

provide proper registration. Varying the threshold setting will show which lines have the greatest sensitivity, because differences in the data readings will appear (Fig. 10-4). In this case, one bit appears to be sensitive to noise caused by several lines all switching simultaneously. An oscilloscope will offer the additional analysis capability desired once the problem line or lines have been located.

Consider next the "lost in software" problem. This can also be called a "where is it?" type of problem. Typically, this is the case where the machine is executing software, but it is not executing the software correctly. The machine is stuck off in a loop somewhere, and the first problem faced by the designer or troubleshooter is to find out where in the software it is stuck, why it is stuck, and how to bring it back. This is a relatively simple problem to solve with a logic state analyzer, because the designer can simply turn off the trigger switches and allow the machine to free-run and acquire whatever data are present on the address lines. It is then a simple matter to map the data, and ascertain in what area of code the machine is operating directly from the coordinates of the map. Even easier, he can press the single-shot switch and arbitrarily grab sixteen words of address data to compare with his program listing. It is particularly useful to use a Hewlett-Packard 1600S to allow the troubleshooter to see not only the address or program sequences that are being executed, but the associated data or instruction words as they actually appear in response to the addressing.

Finding software errors or data-dependent errors is much quicker and more accurate, because there is no need to speculate on what sort of instruction might be causing the erroneous state flow; the actual instruction is recorded. Having located where in his code the machine is stuck, the troubleshooter can select meaningful trigger words to either "walk" his way through to the point at which exiting from the loop is expected, or he can use the END display trigger mode to walk backwards to the origin or entry points of the loop if he so desires. Usually, finding where the processor came from is more challenging than finding out where it is.

▶

Figure 10-4 (a) An H-P 1600S system is used to measure data on an I/O bus read at two different threshold settings with the resultant data XOR'd in the B-memory position. (b) On the word following the highlighted trigger words in the A-memory, one of the bits dipped momentarily to a zero. As the threshold setting was adjusted high enough, the excess ringing caused by the simultaneous switching of a large number of bits on the same clock showed up as a logic error. *(Courtesy, Hewlett-Packard.)*

```
0111  1010  0100  0000    0000  0000  0000  0000
0111  1010  0000  0000    0000  0000  0000  0000

1100  1011  0001  0001    0000  0000  0000  0000
1100  1011  0101  0101    0000  0000  0000  0000

1001  0100  1111  1110    0000  0000  0000  0000
1000  0101  1111  1111    0000  0000  0000  0000

0011  0100  1010  0110    0000  0000  0000  0000
0111  1010  0100  1100    0000  0000  0000  0000

0111  1000  1000  0100    0000  0000  0000  0000
0111  0110  0010  0100    0000  0000  0000  0000

0111  0100  1010  0000    0000  0000  0000  0000
0111  1010  0000  0000    0000  0000  0000  0000

0111  1010  0000  0000    0000  0000  0000  0000
0111  1010  0000  0000    0000  0000  0000  0000

0111  1010  0000  0000    0000  0000  0000  0000
0111  1010  0000  0000    0000  0000  0000  0000
```

(a)

```
0111  1010  0100  0000    0000  0000  0000  0000
0111  1010  0000  0000    0000  0000  0000  0000

1100  1011  0001  0001    0000  0000  0000  0000
1100  1011  0101  0101    0000  0000  0000  0000

1001  0100  1111  1110    0000  0000  0000  0000
1000  0101  1111  1111    0000  0000  0000  0000

0011  0100  1010  0010    0000  0000  0000  0100
0111  1010  0100  1100    0000  0000  0000  0000

0111  1000  1000  0100    0000  0000  0000  0000
0111  0110  0010  0100    0000  0000  0000  0000

0111  0100  1010  0000    0000  0000  0000  0000
0111  1010  0000  0000    0000  0000  0000  0000

0111  1010  0000  0000    0000  0000  0000  0000
0111  1010  0000  0000    0000  0000  0000  0000

0111  1010  0000  0000    0000  0000  0000  0000
0111  1010  0000  0000    0000  0000  0000  0000
```

(b)

The causes of failure to exit from loops or causes for arriving in strange areas of code are, of course, manifold. Such things as software errors, flag signals that fail to be set or reset, or fail to arrive on time, noise, any number of things can cause the machine to go crashing off into the software "weeds." The reason that a logic state analyzer can help solve these problems quickly is that it does not require that the user know anything whatsoever about the dimensions of the problem to make the first measurement. He simply grabs arbitrarily 16 words of data, or looks at the map and asks, "Where is it?" The logic analyzer tells him exactly. Solving the problem from that point may or may not be straightforward; it may require thoughtful analysis to determine the cause of the error, but the logic state analyzer will certainly point out the location to look at next for that cause.

The data domain and the time domain are often inseparable. Another fairly common problem calling for real-time analysis is one mentioned previously, one where a flag signal is being incorrectly read. A generalized example is shown in Fig. 10–5. At program address 1002, the external I/O signal must be "set" in order for the machine to correctly exit the loop. An oscilloscope or logic probe verifies that the flag signal is present, but the machine fails to recognize it. The problem is one of timing, and to analyze it properly the oscilloscope must be synchronized to the proper moment in state time, not triggered from the detection of the signal itself. One of the key features of a logic state analyzer is the trigger output. Two trigger outputs are provided: one, when a preset pattern trigger occurs and is recognized, a second one when the delay is counted down. In this case, the pattern trigger output can be used directly, because it is known exactly at which program address, 1002, the flag signal is to be read. Externally triggering the oscilloscope on the occurrence of that address enables examination of the flag signal synchronized in time to the software. Once that a state-time reference is established, determining whether it is early or late is easy.

Note the need again for time domain analysis together with state analysis. The two are quite often inseparable and form the basis for instrumentation such as the "gold button" oscilloscope from Hewlett-Packard that provides single-button selection of either time domain display or logic state analysis in a system. As a final example, consider the problems of dual clock troubleshooting across an I/O port. In this example, an IC tester has been designed that utilizes a 4-bit microprocessor (Intel 4004) to form a test station under handshake control from an 8-bit supervisory microprocessor

General Considerations · 241

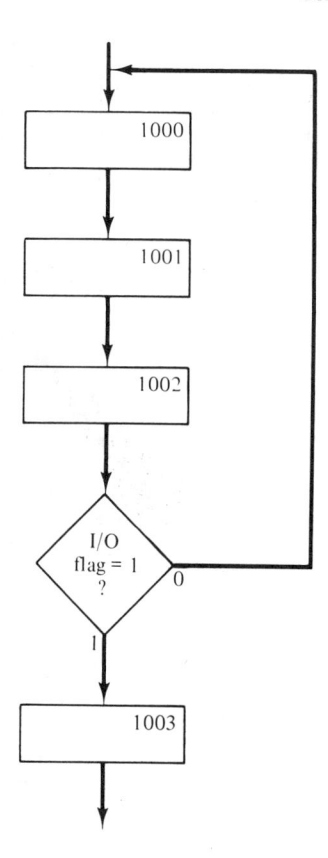

```
    ┌─────┐
    │1000 │←─────┐
    └──┬──┘      │
       ↓         │
    ┌─────┐      │
    │1001 │      │
    └──┬──┘      │
       ↓         │
    ┌─────┐      │
    │1002 │      │
    └──┬──┘      │
       ↓         │
      ╱ I/O ╲    │
     ╱ flag=1╲ 0 │
     ╲   ?   ╱───┘
      ╲     ╱
        │ 1
        ↓
    ┌─────┐
    │1003 │
    └──┬──┘
       ↓
```

Figure 10-5 Illustration of a typical loop problem. Failure to exit from the loop is caused by failure to detect or properly read the "flag" signal at program address to 1002. An oscilloscope can be used to examine the "flag" line synchronized in time to address 1002 via the trigger output of a logic state analyzer. *(Courtesy, Hewlett-Packard.)*

(Motorola 6800). The supervisor instructs the test station to execute a complete functional check of the truth table of a quad 2-input NAND gate. What should be verified under real-time conditions is that the system operates correctly, that the hardware responds correctly across the I/O as a unique function of the software.

Again, a H-P 1600S system is used to provide the capability of reading data on two independent clocks (Fig. 10-6). Memory A displays 8 bits of the program sequence plus the 8 bit instructions of the remote unit, and memory B displays the execution of the func-

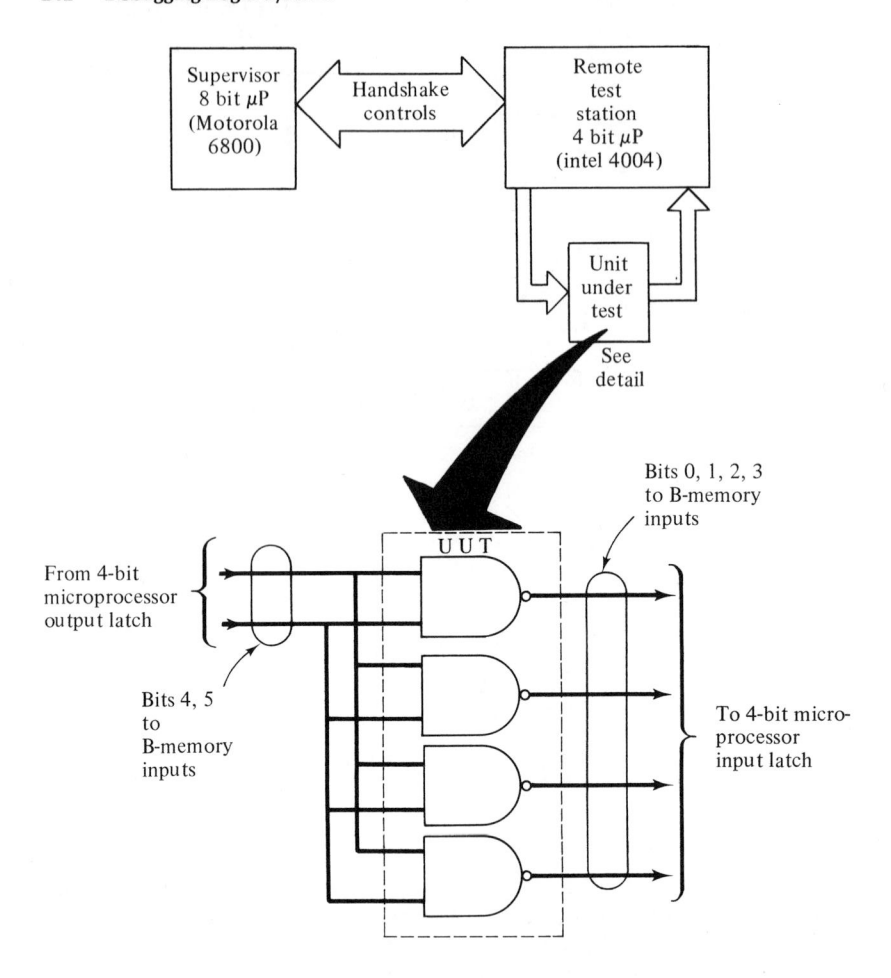

Figure 10-6 Dual clock measurement using the H-P 1600S to view simultaneously on one screen both the software of the microprocessor unit (the program addressed) and the results of the functional test at the I/O of the test station. *(Courtesy, Hewlett-Packard.)*

tional tests by the test station. The two displays are synchronized together (Fig. 10-7) at the unique program step, hex address 22E2 via the bus trigger. The B-memory clearly shows the four possible conditions on channels 4 and 5 and the four outputs correct for the NAND function read on channels 0 through 3. Note only four lines (clocks) of data are read into the B-memory because the test was executed single-shot.

```
0010  0010  1110  0010      11  0000
0010  0011  1110  1010      00  1111
0010  0100  1111  0100      10  1111
0010  0101  0001  1100      01  1111

0010  0110  0100  0011
0010  0111  1111  0000
0010  1000  1111  1010
0010  1001  1111  0101
0010  1010  1110  0010
0010  1011  1110  1010
0010  1100  1111  0100
0010  1101  0001  1100
0010  1110  0100  0011
0010  1111  0000  0000
0011  0000  1111  0000
0011  0001  1111  1010
```

Figure 10–7 An H-P 1600S in a dual clock measurement. The A-memory is showing the software (8 bits of program count + 8 bit instruction words) of the remote microprocessor (Intel 4004), while the B-memory shows the I/O data at the test station testing a quad 2-input NAND gate. *(Courtesy, Hewlett-Packard.)*

The last example, one involving data transfer across an I/O, executed in real time under dual clock conditions, measured without interference by the measurement tool itself, summarizes in one experiment many of the special abilities of the logic state analyzers. In other examples, more mundane, perhaps, but often troublesome, measurements were illustrated involving parametric as well as logic state flow problems. The designer should be able to identify similar problems he has encountered.

10–2 FUNCTIONAL ANALYSIS OF NATIONAL IMP MICROPROCESSOR SYSTEMS

This discussion, provided by the Hewlett-Packard Co., is intended to assist the user of National Semiconductor Corp. IMP microprocessor devices in real-time analysis of his system in a design environment. This note demonstrates real-time analysis of program flow, triggering on a specific event, and use of the paging technique. The central processing unit (CPU) of the IMP system is configured around the Control Read Only Memory (CROM) device and one (or more) Register and Arithmetic Logic Unit (RALU) device(s), as shown in Fig. 10-8. A CROM provides storage for 100 microinstructions of 23 bits each, program sequencing, subroutine execution, and translation of microinstructions into RALU commands. Each RALU provides 96 bits of storage; four bits in each of seven general registers; a status register; and a 16-word last-in/first-out (LIFO) stack. The RALU also contains provisions for: an arithmetic/logic unit to perform ADD, AND, OR, XOR operations; a shift register; an input/output (I/O) data multiplexer to an external data bus; and a four-bit, time-multiplexed command bus for RALU control.

Both CROM and RALU chips are constructed using standard P-channel enhancement-mode silicon-gate technology, and operate on +5-V and −12-V power supplies with four-phase nonoverlapping clock signals. Signals interfacing CROM and RALU chips are MOS logic level, while signals interfacing the rest of the system are TTL logic level. Consider next the pin assignments depicted in Fig. 10-9. Pin 1 is for jump connection input from the external jump condition multiplexer. Pins 2-5 and 7-10 are for data input lines and carry instructions to the CROM instruction register. Pin 11 is an enable control that provides logic "1" when branch to instruction fetch routine occurs. It responds to logic "1" input by executing "branch to instruction fetch routine." Pin 12 is the +5 V input. Pins 13-16 are for four-phase nonoverlapping clock signals. Pins 17-20 provide command bits to RALU.

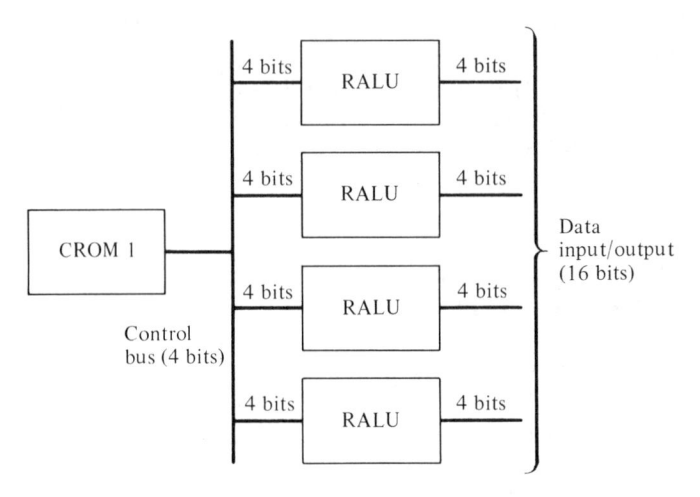

Figure 10-8 Central processing unit (CPU) configuration. *(Courtesy, Hewlett-Packard.)*

Pin 21 is for low-order carry/shift to and from RALU. Pin 22 is for high-order carry/shift to and from RALU. Pin 23 is a flag enable control output used to set/reset flags addressed by data inputs. Pin assignments depicted in Fig. 10–8 are as follows: Pins 1, 2, 22, and 23 are for four-phase nonoverlapping clock signals. Pin 3 is a stack "full" output that goes true when the bottom word of stack is nonzero at the start of the previous cycle. Pins 4, 5, 7, 17 transfer data between RALU and memory or peripheral devices. Pin 6, "result" bus equals zero. The logic level goes to "0" level when R-bus contains all zeroes. Pin 9, the save/restore line, provides a means of modifying status flags over the data bus. Pin 10, SININ, accomplishes sign extension. Pins 11 and 14 carry input (CSHO) and carry output (CSH3) used primarily for transfer of carry and shift informa-

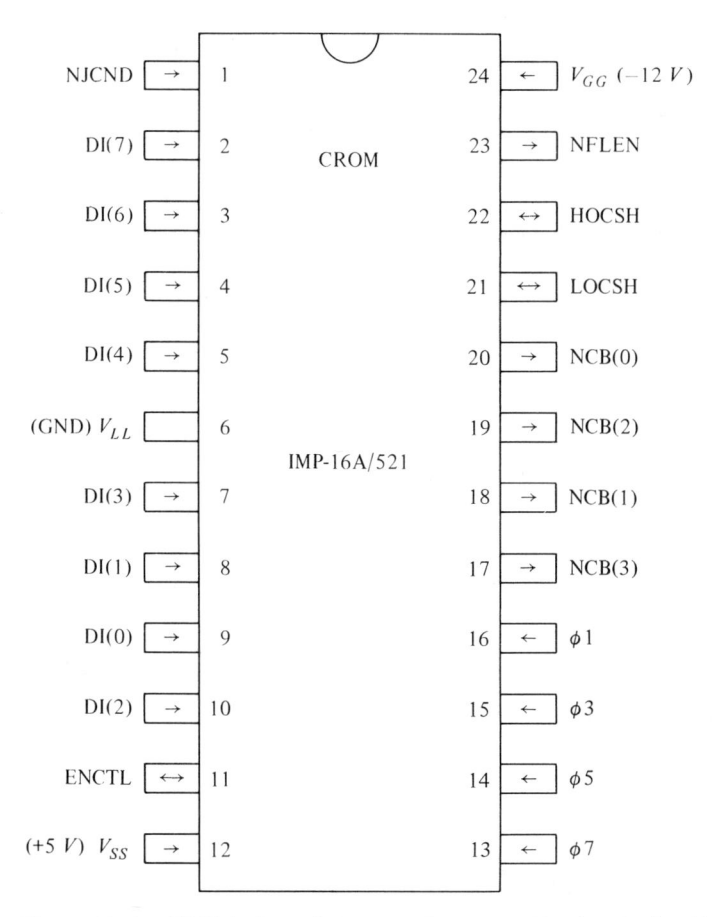

Left Pin		#	Chip	#		Right Pin
NJCND	→	1	CROM	24	←	V_{GG} (−12 V)
DI(7)	→	2		23	→	NFLEN
DI(6)	→	3		22	↔	HOCSH
DI(5)	→	4		21	↔	LOCSH
DI(4)	→	5		20	→	NCB(0)
(GND) V_{LL}		6	IMP-16A/521	19	→	NCB(2)
DI(3)	→	7		18	→	NCB(1)
DI(1)	→	8		17	→	NCB(3)
DI(0)	→	9		16	←	$\phi 1$
DI(2)	→	10		15	←	$\phi 3$
ENCTL	↔	11		14	←	$\phi 5$
(+5 V) V_{SS}	→	12		13	←	$\phi 7$

Figure 10–9 CROM pin assignments. *(Courtesy, Hewlett-Packard.)*

tion between RALU's or between RALU and CROM. Pin 13, select flag input, is used to select carry or overflow and to determine whether the link is included in shift operations. Pin 15, carry/overflow, provides output signal indicating the state of Carry flag or Overflow flag as selected by the select input. Pin 16, flag output, indicates the state of the general-purpose status flag. Pins 18–21 command inputs to RALU.

Consider next the probe connections. All information on address and data acquisition explained below applies specifically to the National Semiconductor IMP 16C (Integrated Microprocessor). However, the procedures outlined are also valid for other microprocessor systems, provided that they are configured to meet the requirements for using a single CROM (IMP-16A/521) chip and four RALU (IMP-16A/520) chips, with the appropriate auxiliary circuits and instruction set (see Figs. 10–9 and 10–10). A system that will not "come up" can frequently be debugged by monitoring address flow alone. Since address data in the IMP system are time-multiplexed with data, the address must be latched by external devices such as National Semiconductor part DM8551. Therefore, the best place to secure address information is at the memory address control lines of the RALU's. The H-P 1600A data probes connected as follows will provide a display of the activity on the address lines (see Fig. 10–11).

To set the controls, turn the power on and set the logic state analyzer controls as follows: Display Mode, Table A; Sample Mode, REPET. Note that if the program is not looping or cycling through the selected address, select SGL, press RESET, and start the system. The first time that the system passes through the trigger point, the display will be generated and stored. Next, set the Trigger Mode: NORM/ARM, NORM; LOCAL/BUS, LOCAL; OFF/WORD, WORD; Start Display, On; Clock, ⎤ , Threshold, TTL. If it is desired to trigger on a signal line that interfaces CROM with RALU, it may be necessary to switch to VAR and adjust to the proper MOS level. All Other Pushbuttons are set to their Out Position; Display Time, ccw; Column Blanking, ccw; Qualifier Q0 LO, and Q1 HI; Trigger Word Switches, are set to Address on which a trigger is required.

Next, consider the display interpretation. When power is first applied to the microprocessor system, all RALU registers, flags, and the stack are cleared to zero. In this illustration, system response to a program for clearing all accumulators to zero is considered. Proper operation is confirmed by a comparison between real-time state analysis (Fig. 10–12) and the assembler listing output (Fig. 10–12).

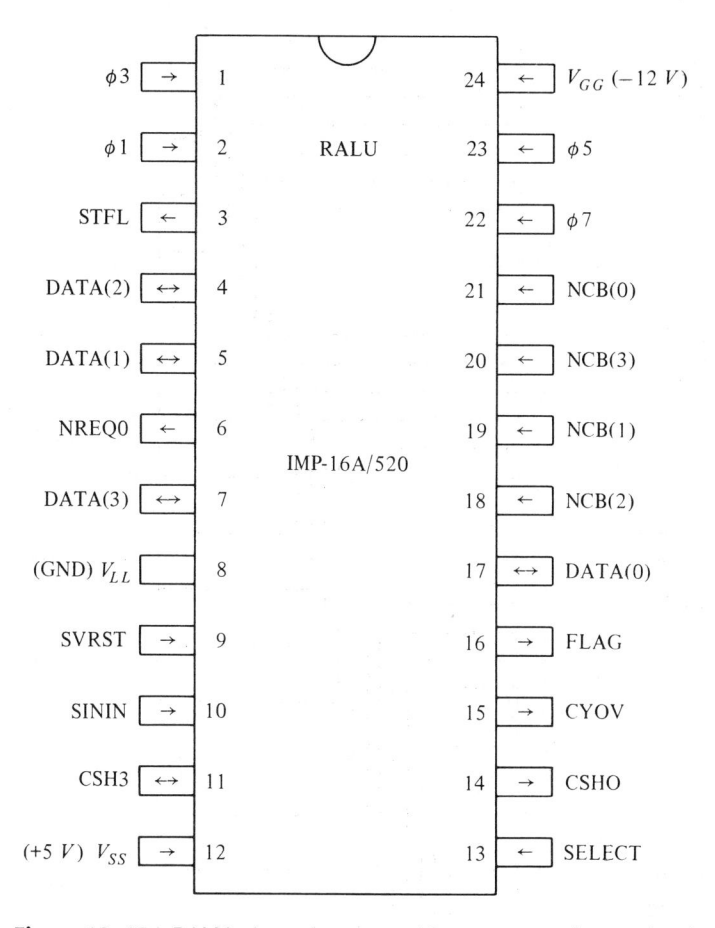

Functional Analysis of National IMP Microprocessor Systems · 247

```
     φ3  →  │ 1              24 │ ←  V_GG (-12 V)
     φ1  →  │ 2     RALU     23 │ ←  φ5
    STFL  ← │ 3              22 │ ←  φ7
 DATA(2)  ↔ │ 4              21 │ ←  NCB(0)
 DATA(1)  ↔ │ 5              20 │ ←  NCB(3)
   NREQ0  ← │ 6              19 │ ←  NCB(1)
            │    IMP-16A/520
 DATA(3)  ↔ │ 7              18 │ ←  NCB(2)
(GND) V_LL  │ 8              17 │ ↔  DATA(0)
   SVRST  → │ 9              16 │ →  FLAG
   SININ  → │ 10             15 │ →  CYOV
    CSH3  ↔ │ 11             14 │ →  CSHO
(+5 V) V_SS → │ 12            13 │ ←  SELECT
```

Figure 10–10 RALU pin assignments. *(Courtesy, Hewlett-Packard.)*

In this illustration, triggering occurs on the first address in the actual program sequence, 0079, to display each subsequent address. Note that at address 007C, the operation code includes a subtract instruction affecting accumulator 1. Line 5 of the table display verifies that the loop address 0078 does indeed take place, and that the instruction is being performed. At the conclusion of the subtract operation, we would expect the system to reenter the program at address 007D. Line 6 of the table verifies system operation by displaying address 007D as expected in a similar manner, each instruction in the routine being shown to have been properly executed.

To view addresses following the last displayed address, simply

248 · Debugging Logic Systems

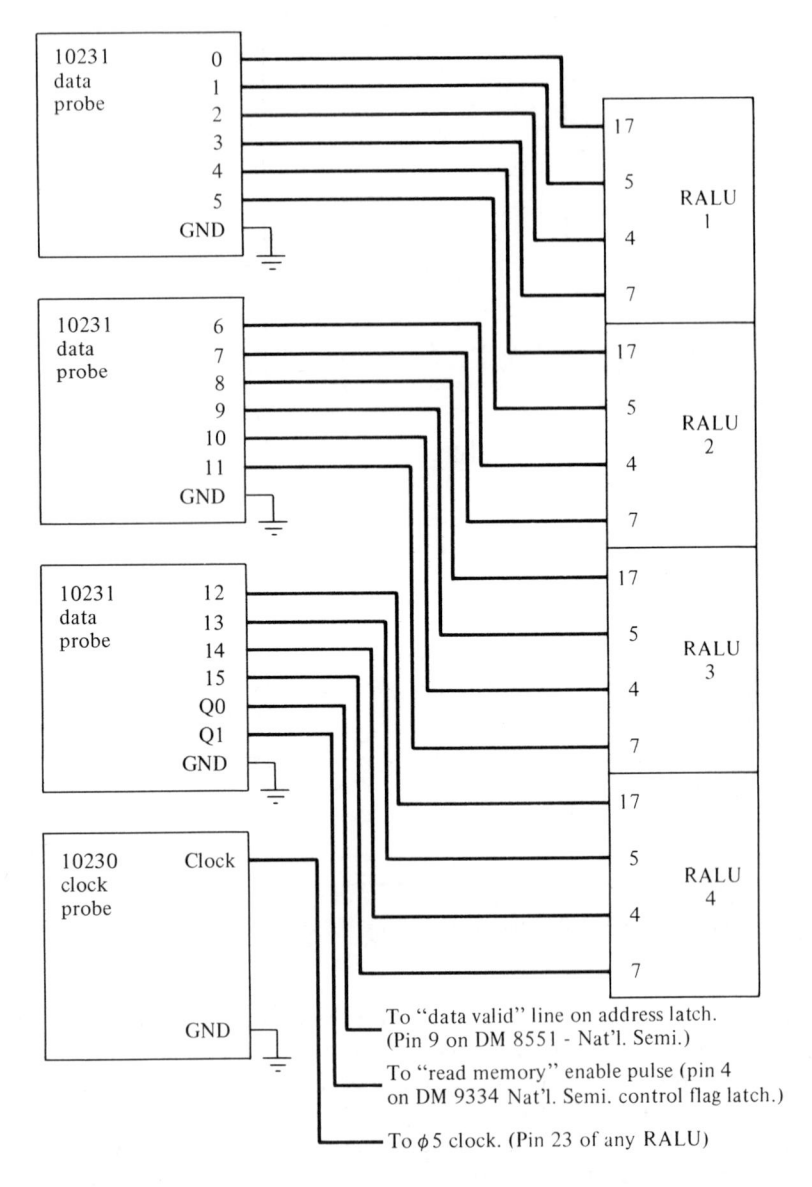

Figure 10-11 Test connection arrangement. *(Courtesy, Hewlett-Packard.)*

set the Trigger Word switches to match the address displayed in line 16. This address then becomes the trigger words in line 1 with the next 15 addresses displayed on lines 2 through 16. If the designer

Functional Analysis of National IMP Microprocessor Systems · 249

			ACCUMULATOR ADDRESS
	0078	A	
			AC 0
5	0000	A	AC 1
6	0001	A	AC 2
7	0002	A	AC 3
8	0003	A	
9	0079	4C02 A	SQR: L1 AC0, X'0002
10	007A	3281 A	RCPY AC0, AC2
11	007B	3981 A	RCPY AC2, AC1
12	007C	D478 A	LOOP: SUB, AC1, ONE
13	007D	3800 A	ADDR: RADD AC2, AC0
14	007E	49FF A	A1SZ AC1, -1
15	007F	207D A	JMP ADDR
16	0080	0600 A	ROUT AC0
17	0081	4C0F A	L1 AC0, X'000F
18	0082	3882 A	RXOR AC2, AC0
19	0083	11FS A	BOC 1, SQR
20	0084	C878 A	ADD AC2, AC0
21	0085	3881 A	RCPY AC2, AC0
22	0086	3181 A	RCPY AC0, AC1
23	0087	207C A	JMP LOOP
24		007C A	•END LOOP

(b)

```
 1  0000 0000 0111 1001
 2  0000 0000 0111 1010
 3  0000 0000 0111 1011
 4  0000 0000 0111 1100
 5  0000 0000 0111 1000
 6  0000 0000 0111 1101
 7  0000 0000 0111 1110
 8  0000 0000 1000 0000
 9  0000 0000 1000 0001
10  0000 0000 1000 0010
    0000 0000 1000 0011
    0000 0000 1000 0100
    0000 0000 0111 1000
    0000 0000 1000 0101
    0000 0000 1000 0110
    0000 0000 1000 0111
Line
```

(a)

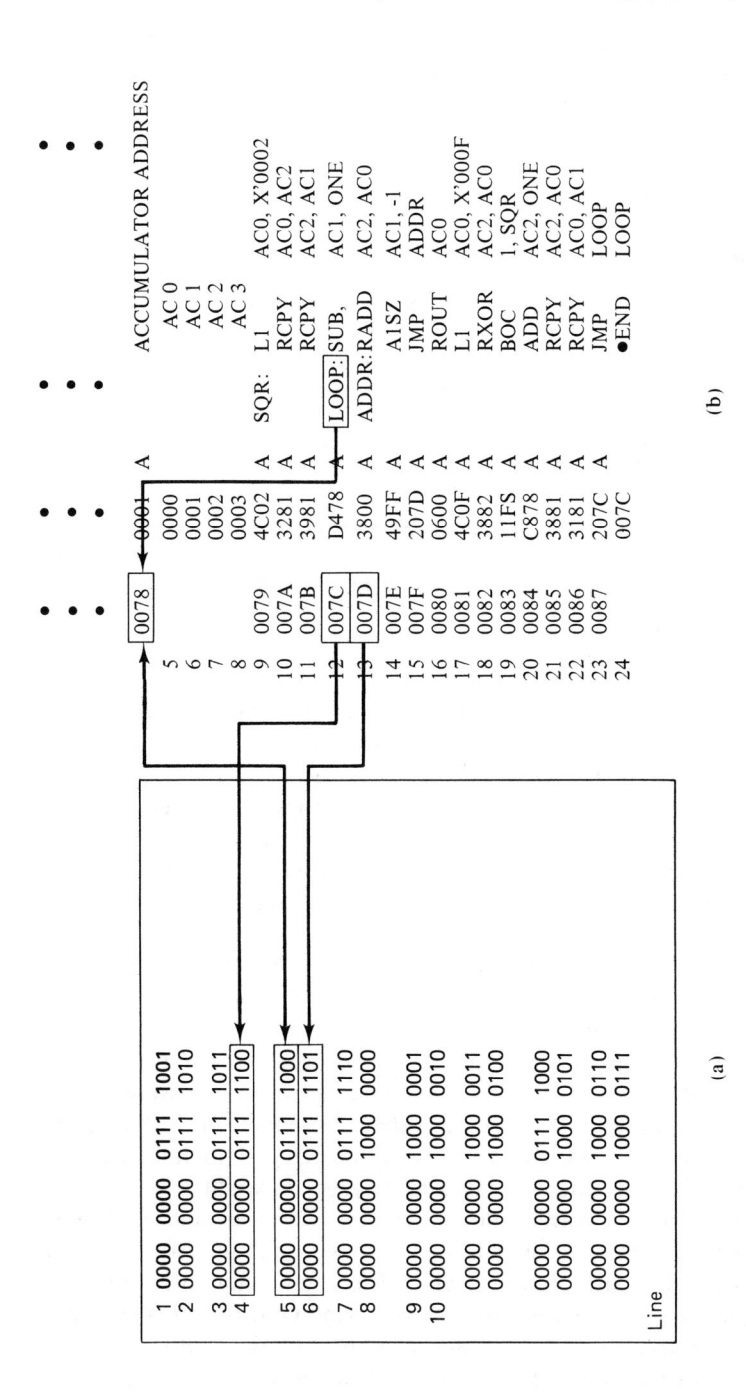

Figure 10–12 System response to routine for clearing accumulators to zero. *(Courtesy, Hewlett-Packard.)*

250 · Debugging Logic Systems

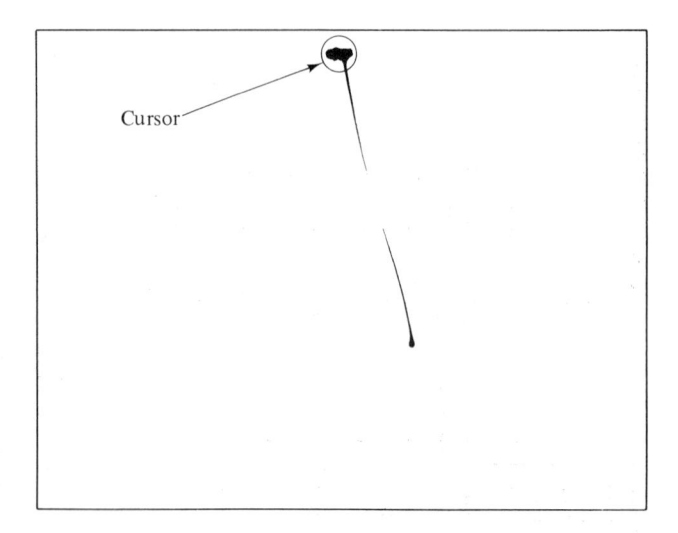

Figure 10–13 Map display shows entire system activity. *(Courtesy, Hewlett-Packard.)*

wishes to retain the original trigger point, an alternate technique is to use digital display and set the thumbwheels to 00015, which provides the same display. Next, it is instructive to consider the map. If a tabular display is not achieved, it means that the system did not access the selected address and the No Trigger light will be on. To find where the system is residing in the program, switch to MAP NORM. Using the Trigger Word switches, move the cursor to encircle one of the dots, as shown in Fig. 10–13. Switch to MAP EXP and make final positioning of the cursor. The No Trigger light will now go out, and switching back to Table A displays the 16 addresses around that point.

When program deviations are found, the reason may be as simple as a program error, or as complicated as a hardware failure on the data bus, the control bus, or other command lines. Additional input channels now become very desirable. By combining the H-P 1600A and 1607A, the display and trigger capability can be expanded to 32 bits in width, allowing 16-bit addresses and 16 bits of data or active command signals to be viewed simultaneously. With regard to display interpretation of address and data lines, observe the sample program in Fig. 10–14. By displaying both address and data, it is possible to confirm exact system operation with respect to the routine for clearing the accumulators to zero.

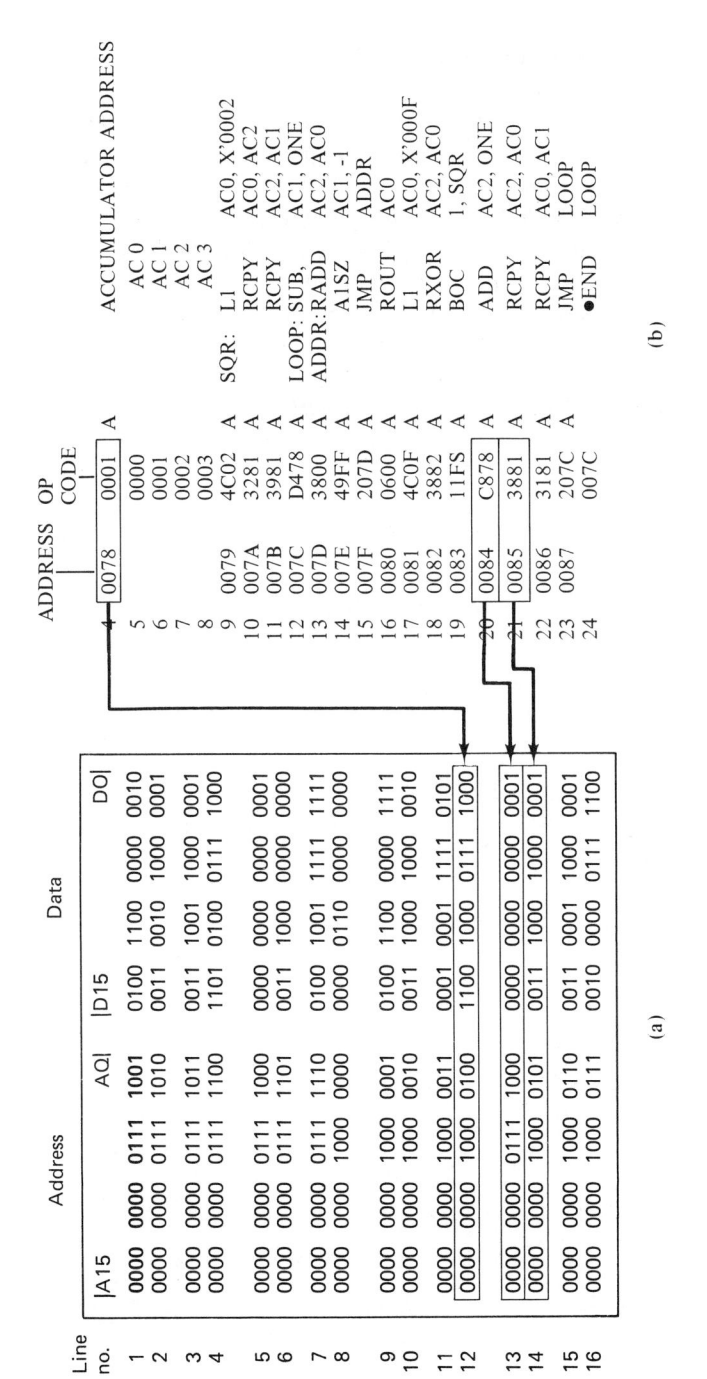

Functional Analysis of National IMP Microprocessor Systems · 251

Line no.	Address A15 ... A0	Data D15 ... D0		ADDRESS	OP CODE		ACCUMULATOR ADDRESS
				0078	0001 A		AC 0
1	0000 0000 0111 1001	0100 1100 0000 0010					AC 1
2	0000 0000 0111 1010	0011 0010 1000 0001		0079	0000		AC 2
				007A	0001		AC 3
3	0000 0000 0111 1011	0011 1001 1000 0001		007B	0002		
4	0000 0000 0111 1100	1101 0100 0111 1000		007C	0003		
5	0000 0000 0111 0111	0000 1000 0000 0001		0079	4C02 A	SQR: L1	AC0, X'0002
6	0000 0000 0111 1101	0011 1000 0000 0000		007A	3281 A	RCPY	AC0, AC2
7	0000 0000 0111 1110	0100 1001 1111 1111		007B	3981 A	RCPY	AC2, AC1
8	0000 0000 0111 1111	0010 0000 0111 1101		007C	D478 A	LOOP: SUB,	AC1, ONE
				007D	3800 A	ADDR:RADD	AC2, AC0
9	0000 0000 1000 0000	0100 0000 0000 0000		007E	49FF A	A1SZ	AC1, -1
10	0000 0000 1000 0001	0100 1100 0000 1111		007F	207D A	JMP	ADDR
11	0000 0000 1000 0010	0011 1000 1000 0010		0080	0600 A	ROUT	AC0
12	0000 0000 1000 0011	0001 0001 1111 0101		0081	4C0F A	L1	AC0, X'000F
				0082	3882 A	RXOR	AC2, AC0
13	0000 0000 0111 1000	1100 1000 0000 0001		0083	11FS A	BOC	1, SQR
14	0000 0000 1000 0100	0011 1000 1000 0001		0084	C878 A	ADD	AC2, ONE
				0085	3881 A	RCPY	AC2, AC0
15	0000 0000 1000 0101	0011 0001 1000 0001		0086	3181 A	RCPY	AC0, AC1
16	0000 0000 1000 0110	0010 0000 0111 1100		0087	207C A	JMP	LOOP
	0000 0000 1000 0111	0000 0000 0111 1100				●END	LOOP

(a)   (b)

Figure 10–14 System response to routine for clearing accumulators to zero on address and data lines. *(Courtesy, Hewlett-Packard.)*

252 · Debugging Logic Systems

For example, line 12 of the state display in Fig. 10-14 shows an operating code that requires an add operation in accumulator 2. Line 13 of the state display verifies that the accumulators are indeed addressed, so the add instruction may be performed. Upon completion of the add instruction, we would anticipate a return to the next address of the program. That this is indeed accomplished is verified by line 14 of the state display, where address 0085 and operating code 3881 are displayed. In a similar manner, each line of the display can be examined to reveal exact program operation. To observe the action of other control lines, connect the Q1 probe of the H-P 1607A to the desired point in the circuit. Setting the qualifier Q1 switch to the HI or LO position, as required by the selected control line state, one can view the effect of the command line on the display.

APPENDIX I

Color Codes

Resistor color codes

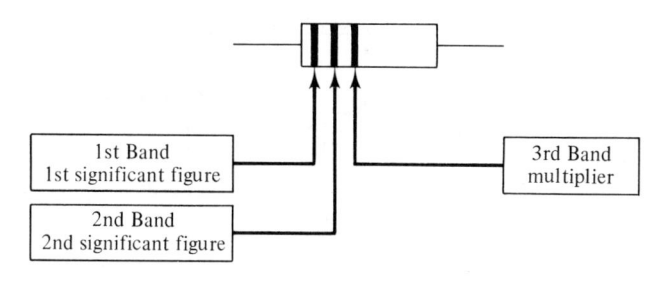

1st Band 1st significant figure		3rd Band multiplier
2nd Band 2nd significant figure		

3-band code for ±20 percent resistors only

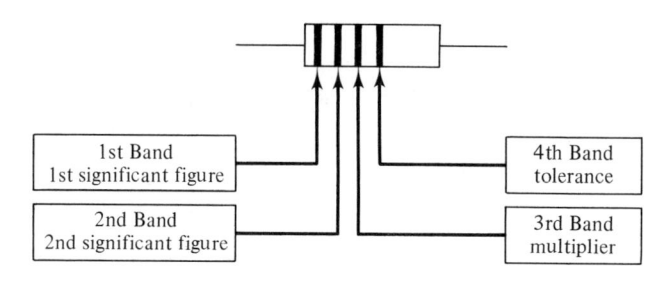

1st Band 1st significant figure		4th Band tolerance
2nd Band 2nd significant figure		3rd Band multiplier

4-band code for resistors with ±1 percent to ±10 percent tolerance

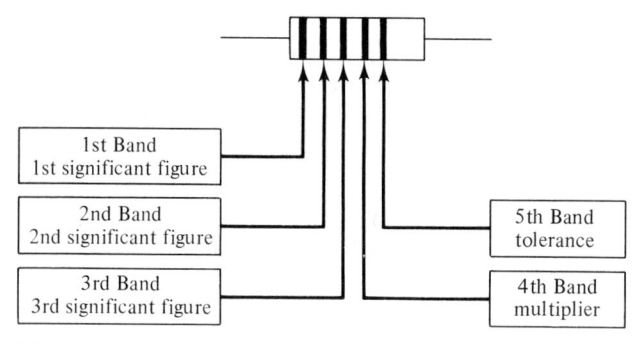

1st Band 1st significant figure		
2nd Band 2nd significant figure		5th Band tolerance
3rd Band 3rd significant figure		4th Band multiplier

5-band code for resistors with ±1 percent to ±10 percent tolerance

Appendix I: Color Codes

First band — 1st digit

Color	Digit
Black	0
Brown	1
Red	2
Orange	3
Yellow	4
Green	5
Blue	6
Violet	7
Gray	8
White	9

Third band — multiplier

Color	Multiplier
Black	1
Brown	10
Red	100
Orange	1,000
Yellow	10,000
Green	100,000
Blue	1,000,000
Silver	0.01
Gold	0.1

Fifth band* reliability level (percent per 1,000 hours)

Color	Level
Brown	M = 1.0%
Red	P = 0.1%
Orange	R = 0.01%
Yellow	S = 0.001%

*MIL-R-J9008 resistors only

Second band — 2nd digit

Color	Digit
Black	0
Brown	1
Red	2
Orange	3
Yellow	4
Green	5
Blue	6
Violet	7
Gray	8
White	9

Fourth band* resistance tolerance

Color	Tolerance
Silver	±10%
Gold	± 5%

*No band ±20%

5-band code for resistors with ±1 percent to ±10 percent tolerance (fifth band reliability level)

Appendix I: Color Codes · 255

Capacitor color codes

Mica capacitor color code

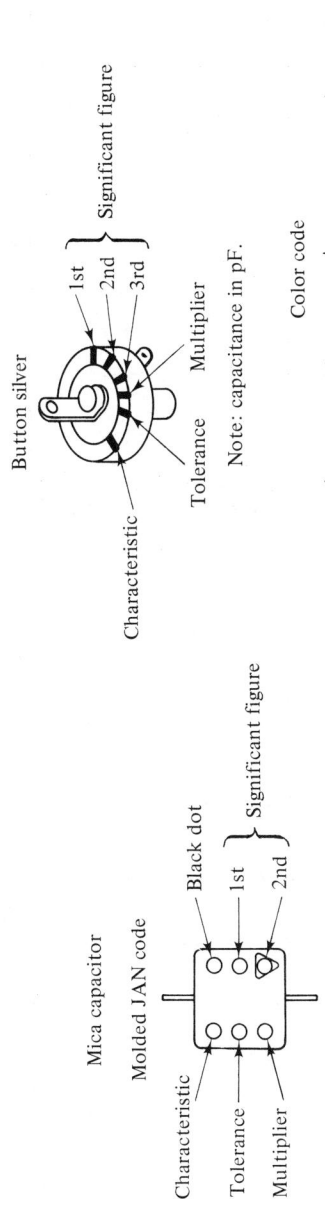

Button silver

1st, 2nd, 3rd — Significant figure
Tolerance
Multiplier
Characteristic

Note: capacitance in pF.

Color code

Color	Significant figure	Multiplier	Tolerance %	Characteristic
Black	0	1	20	A
Brown	1	10	—	B
Red	2	10^2	2	C
Orange	3	10^3	3(RETMA)	D
Yellow	4	10^4	—	E
Green	5	—	5(RETMA)	F (JAN)
Blue	6	—	—	G (JAN)
Violet	7	—	—	—
Gray	8	—	—	I (RETMA)
White	9	—	—	J (RETMA)
Gold	—	0.1	5(JAN)	—
Silver	—	0.01	10	—
None	—	—	20(RETMA)	—

Mica capacitor
Molded JAN code

Black dot
1st, 2nd — Significant figure

Characteristic
Tolerance
Multiplier

Retma code

White dot
1st, 2nd — Significant figure

Characteristic
Tolerance
Multiplier

Note: capacitance in pF. If both rows of dots are not on one face, rotate capacitor about the axis of its leads and read second row on side or rear.

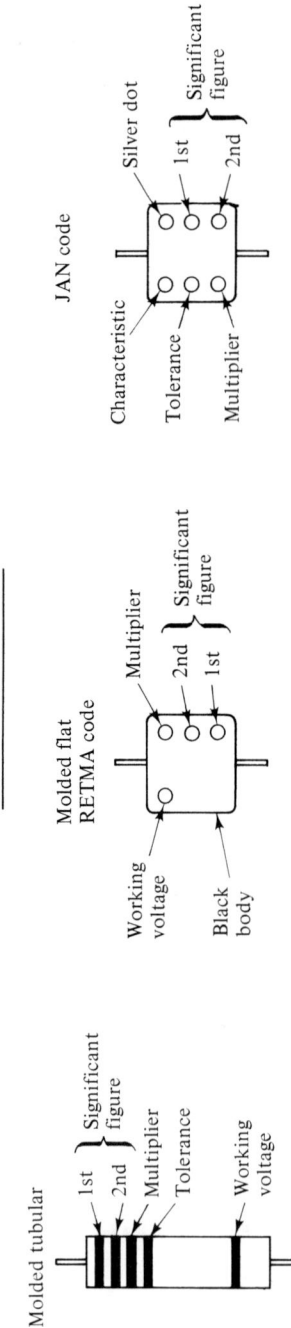

256 · Appendix I: Color Codes

Paper capacitor color code

Molded tubular

1st ⎫ Significant
2nd ⎭ figure
Multiplier
Tolerance
Working voltage

Color	Significant figure	Multiplier	Tolerance %
Black	0	1	20
Brown	1	10	—
Red	2	10^2	—
Orange	3	10^3	30
Yellow	4	10^4	40
Green	5	10^5	5
Blue	6	10^6	—
Violet	7	—	—
Gray	8	—	—
White	9	—	10
Gold	—	1	—

Note: capacitance in picofarads. Working voltage coded in terms of hundreds of volts. Ratings over 900 V expressed in two-band voltage code.

Molded flat RETMA code

Multiplier
2nd ⎫ Significant
1st ⎭ figure
Working voltage
Black body

JAN code

Silver dot
1st ⎫ Significant
2nd ⎭ figure
Characteristic
Tolerance
Multiplier

Color	Significant figure	Multiplier	Tolerance %	Characteristic
Black	0	1	20	A
Brown	1	10	—	B
Red	2	10^2	—	C
Orange	3	10^3	30	D
Yellow	4	10^4	40	E
Green	5	10^5	5	F
Blue	6	10^6	—	G
Violet	7	—	—	—
Gray	8	—	—	—
White	9	—	10	—
Gold	—	1	5	—

Note: capacitance in picofarads. Working voltage coded in terms of hundreds of volts. Voltage ratings over 900 V expressed in two-dot voltage code.

Appendix I: Color Codes · 257

Ceramic capacitor

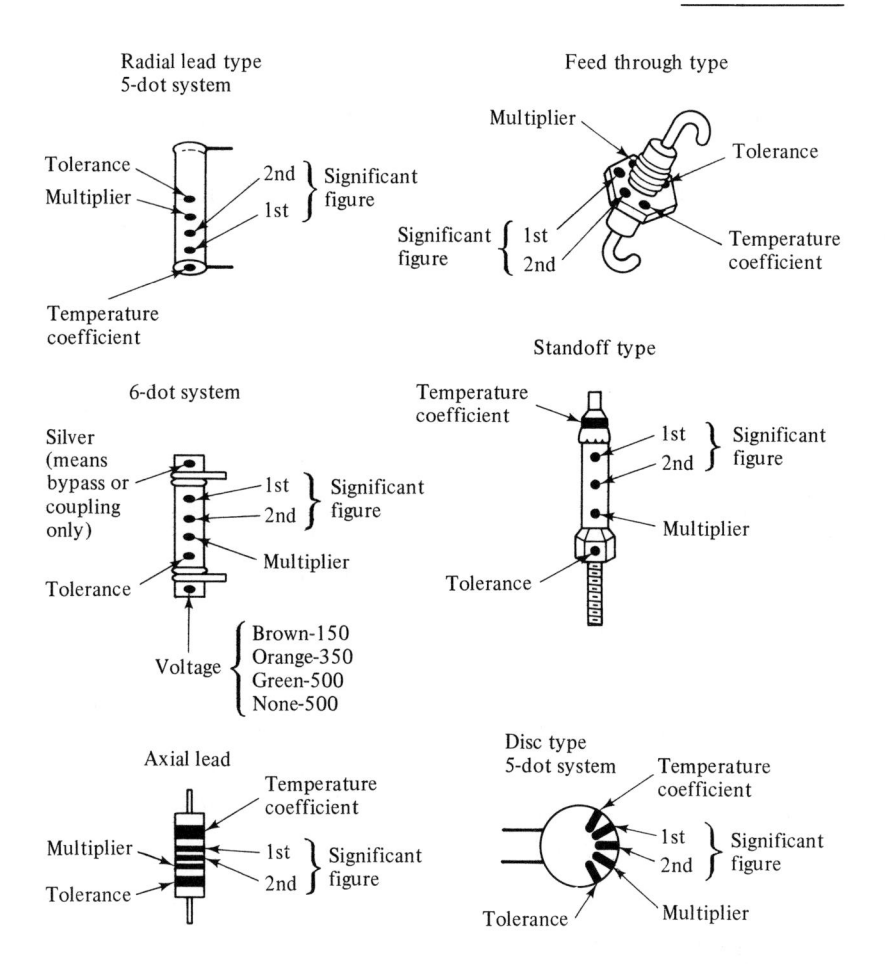

Radial lead type
5-dot system
- Tolerance
- Multiplier
- 2nd } Significant
- 1st } figure
- Temperature coefficient

Feed through type
- Multiplier
- Tolerance
- Significant { 1st
- figure { 2nd
- Temperature coefficient

6-dot system
- Silver (means bypass or coupling only)
- 1st } Significant
- 2nd } figure
- Multiplier
- Tolerance
- Voltage { Brown-150
  Orange-350
  Green-500
  None-500

Standoff type
- Temperature coefficient
- 1st } Significant
- 2nd } figure
- Multiplier
- Tolerance

Axial lead
- Temperature coefficient
- Multiplier
- 1st } Significant
- 2nd } figure
- Tolerance

Disc type
5-dot system
- Temperature coefficient
- 1st } Significant
- 2nd } figure
- Multiplier
- Tolerance

Appendix I: Color Codes

color code

Color	Significant figure	Multiplier	Tolerance A	Tolerance B	Temperature coefficient
Black	0	1	2	20	0
Brown	1	10	0.1	1	−30
Red	2	10^2	−	2	−60
Orange	3	10^3	−	2.5	−150
Yellow	4	10^4	−	−	−220
Green	5	−	5	5	−330
Blue	6	−	−	−	−470
Violet	7	−	−	−	−750
Gray	8	0.01	0.025	−	+30
White	9	0.1	1	10	+120 to −750 (RETMA) +500 to −330 (JAN)
Gold	−	−	−	−	+100
Silver	−	−	−	−	Bypass or coupling

3-dot system

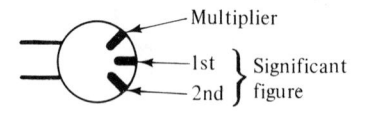

- Multiplier
- 1st } Significant
- 2nd } figure

APPENDIX II

Characteristics of Series- and Parallel-resonant Circuits

Characteristics of series- and parallel-resonant circuits

Quantity	Series circuit	Parallel circuit
At resonance: Reactance ($X_L - X_C$)	Zero; because $X_L = X_C$	Zero; because nonenergy currents are equal
Resonant frequency	$\dfrac{1}{2\pi\sqrt{LC}}$	$\dfrac{1}{2\pi\sqrt{LC}}$
Impedance	Minimum; $Z = R$	Maximum $Z = \dfrac{L}{CR'}$ approx.
I_{line}	Maximum;	Minimum value
I_L	I_{line}	$Q \times I_{line}$
I_C	I_{line}	$Q \times I_{line}$
E_L	$Q \times E_{line}$	E_{line}
E_C	$Q \times E_{line}$	E_{line}
Phase angle between E_{line} and I_{line}	0°	0°
Angle between E_L and E_C	180°	0°
Angle between I_L and I_C	0°	180°
Desired value of Q	10 or more	10 or more
Desired value of R	Low	Low
Highest selectivity	High Q, low R, high $\dfrac{L}{C}$	High Q, low R
When f is greater than f_o: Reactance	Inductive	Capacitive
Phase angle between I_{line} and E_{line}	Lagging current	Leading current
When f is less than f_o: Reactance	Capacitive	Inductive
Phase angle between I_{line} and E_{line}	Leading current	Lagging current

Appendix II: Characteristics of Series- and Parallel-Resonant Circuits

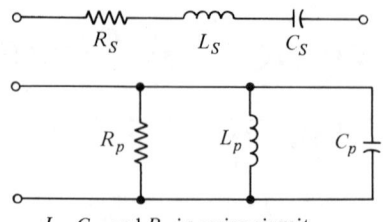

L_s, C_s, and R_s in series circuit.
L_p, C_p, and R_p in parallel circuit.

$$Q = \frac{\omega L_s}{R_s} = \frac{1}{\omega C_s R_s} = \frac{R_p}{\omega L_p} = R_p \omega C_p = \frac{\sqrt{L_s/C_s}}{R_s} = \frac{R_p}{\sqrt{L_p/C_p}}$$

General formulas	Formulas for Q Greater than 10	Formulas for Q Less than 0.1
$R_s = \dfrac{R_p}{1+Q^2}$	$R_s \simeq \dfrac{R_p}{Q^2}$	$R_s \simeq R_p$
$X_s = X_p \dfrac{Q^2}{1+Q^2}$	$X_s \simeq X_p$	$X_s \simeq X_p Q^2$
$L_s = L_p \dfrac{Q^2}{1+Q^2}$	$L_s \simeq L_p$	$L_s \simeq L_p Q^2$
$C_s = C_p \dfrac{1+Q^2}{Q^2}$	$C_s \simeq C_p$	$C_s \simeq \dfrac{C_p}{Q^2}$
$R_p = R_s(1+Q^2)$	$R_p \simeq R_s Q^2$	$R_p \simeq R_s$
$X_p = X_s \dfrac{1+Q^2}{Q^2}$	$X_p \simeq X_s$	$X_p \simeq \dfrac{X_s}{Q^2}$
$L_p = L_s \dfrac{1+Q^2}{Q^2}$	$L_p \simeq L_s$	$L_p \simeq \dfrac{L_s}{Q^2}$
$C_p = C_s \dfrac{Q^2}{1+Q^2}$	$C_p \simeq C_s$	$C_p \simeq C_p Q^2$
$B_L = \dfrac{1}{X_L}$	$B_L = \dfrac{1}{X_L}$	$B_L = \dfrac{1}{X_L}$
$B_C = \dfrac{1}{X_C}$	$B_L = \dfrac{1}{X_L}$	$B_L = \dfrac{1}{X_L}$
$Y = \sqrt{G^2 + B^2}$	$Y = \dfrac{1}{\sqrt{G^2 + B^2}}$	$Y = \sqrt{G^2 + B^2}$

APPENDIX III

Graphical Solution for Inductive and Capacitive Reactances in Parallel

Graphical solution for inductive and capacitive reactances in parallel

($X_L > X_C$)

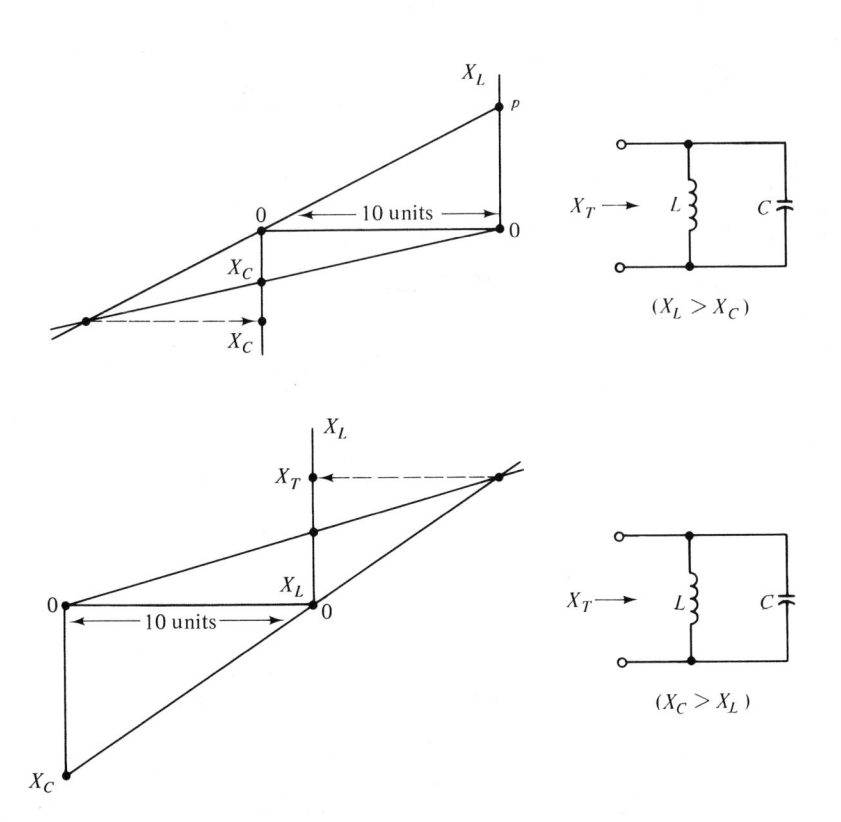

($X_C > X_L$)

APPENDIX IV

Logic Gates and DeMorgan's Equivalents

Gates and De Morgan's equivalents		Table of combinations		
AND	OR	A	B	F
$A \cdot B$	$\overline{A} + \overline{B}$	H	H	H
		H	L	L
		L	H	L
		L	L	L
$\overline{A} \cdot B$	$\overline{A + \overline{B}}$	H	H	L
		H	L	L
		L	H	H
		L	L	L
$A \cdot \overline{B}$	$\overline{\overline{A} + B}$	H	H	L
		H	L	H
		L	H	L
		L	L	L
$\overline{A} \cdot \overline{B}$	$\overline{A + B}$	H	H	L
		H	L	L
		L	H	L
		L	L	H
$\overline{A \cdot B}$	$A + B$	H	H	H
		H	L	H
		L	H	H
		L	L	L
$\overline{A \cdot \overline{B}}$	$\overline{A} + B$	H	H	H
		H	L	L
		L	H	H
		L	L	H
$\overline{\overline{A} \cdot B}$	$A + \overline{B}$	H	H	H
		H	L	H
		L	H	L
		L	L	H
$\overline{\overline{A} \cdot \overline{B}}$	$\overline{A} + \overline{B}$	H	H	L
		H	L	H
		L	H	H
		L	L	H

APPENDIX V

Integrated Injection Logic (I²L)

One of the emerging integrated circuit technologies is Integrated Injection Logic, or I²L. It combines two very desirable properties: simple fabrication and high-speed operation. I²L devices are now the equal of saturated bipolar devices. I²L microprocessors are the beginning of a new family of IC circuits using this technology. An I²L gate is very simple. As shown in Figure A-5-1, the basic gate structure is merely an inverter that is implemented as a multicollector transistor. A current source (a PNP transistor) supplies base drive to the NPN multicollector injection element. The basic logic function is an inverter, and NAND functions can be provided with the basic inverter gate.

One of the logic symbols used for I²L gates is shown in Figure A-5-2. The individual multicollector outputs are shown as separate output lines, with a single input line to the inverter gate. Multiple inputs to the gate are wire- ANDed together as depicted in Figure A-1-2, and the inversion function of the multicollector transistor provides a NAND gate. The multicollectors provide separate output lines. The gate of Figure A-1-2 has a "fan out" of three (capable of driving three additional gates). For a fan out of four or five, additional multicollector outputs would be added to the NON inverter transistor.

The basic I²L gate can be improved by the addition of Schottky

264 · Appendix V: Integrated Injection Logic (I²L)

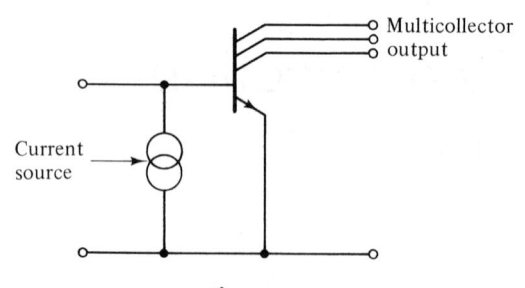

Figure A 5-1

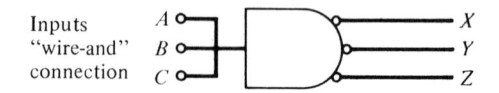

Inputs "wire-and" connection

Figure A 5-2

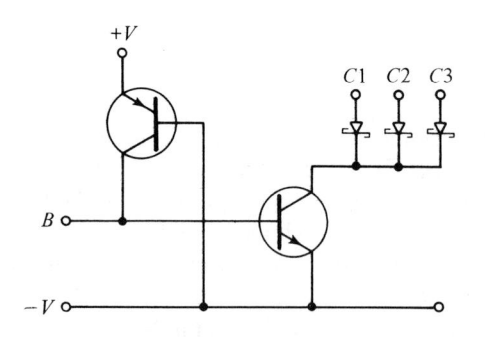

Figure A 5-3

clamp diodes on the output lines, as shown in Figure A-5-3. The addition of the Schottky diodes on the output reduces the voltage swings and increases the operating speed. The simplicity of I²L gate fabrication is illustrated in Figure A-5-4. The two transistors that form the gate are fabricated on a single N-type region. An additional P and N diffusion forms the multicollector transistor and two P diffusions form the PNP current source transistor. The Schottky diodes are formed at the individual multicollectors. The oxide barrier shown in Figure A-1-4 is used to separate the individual sections on the clip. The simplicity of the fabrication assures that the current distribution between the multicollector outputs is uniform. This eliminates the current-hogging condition that plagued early direct-coupled transistor logic systems. The I²L gate eliminates the need for resistors, thus saving chip space and increasing density.

Appendix V: Integrated Injection Logic (I²L) · 265

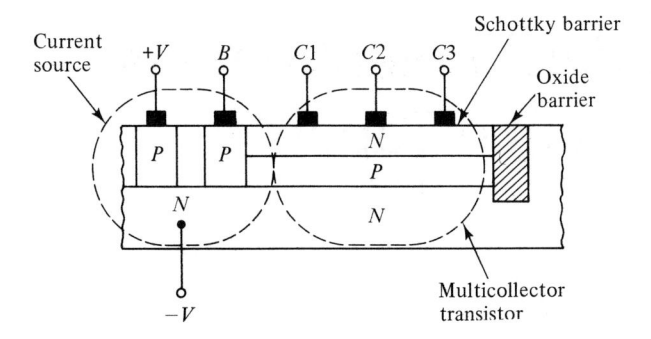

Figure A 5-4

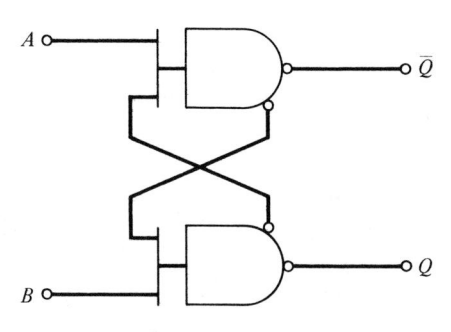

Figure A 5-5

The latch shown in Figure A-5-5 illustrates the simplicity of I²L gate interconnections. Gates 1 and 2 are cross-coupled in the normal inverter fashion and together with the A and B inputs are wire-ANDed at the input. The basic gate structure eliminates connections within the logic element itself and greatly simplifies the inner connection of gate elements. A single circuit can operate over a wide speed range simply by varying the total current into the injection transistor. I²L densities range up to 250 gates per square millimeter. Schottky clamped I²L gates can provide propagation delays in the one to five nano-second range. Without Schottky diodes, propagation delays in the ten nano-second range can be obtained.

APPENDIX VI
Standard Potentiometer Resistance Tapers

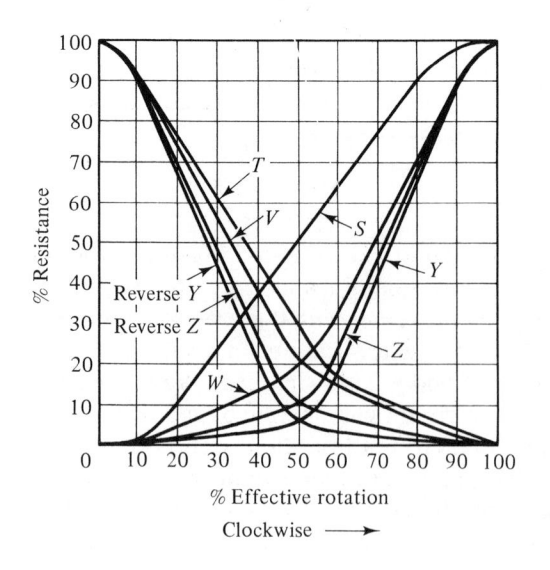

Taper S Straight or uniform resistance change with rotation.
Taper T Right-hand 30% resistance at 50% of C.C.W. rotation.
Taper V Right-hand 20% resistance at 50% of C.C.W. rotation.
Taper W Left-hand 20% resistance at 50% of C.W. rotation.
Taper Z Left-hand (Log. audio) 10% resistance at 50% C.W. rotation.
Taper Y Left-hand 5% resistance at 50% of C.W. rotation.

APPENDIX VII
Resistance
RMA Preferred Values

RESISTANCE
RMA Preferred Values

Commercial tolerances on resistance values are 5, 10, and 20 percent. RMA preferred values are nominal resistance values that avoid duplication of stock within each tolerance range. The nominal resistance values tabulated below can be multiplied or divided by any power of 10.

TOLERANCE

5%	10%	20%
10	10	10
11		
12	12	
13		
15	15	15
16		
18	18	
20		
22	22	22
24		
27	27	
30		
33	33	33
36		
39	39	39
43		
47	47	47
51		
56	56	
62		
68	68	68
75		
82	82	
91		
100	100	100

Index

Absolute maximum, 6
AC
  beta, 96
  resistance, 88
Access time, 231
Accumulator, 246, 252
Active
  filter, 27, 68
  region, 206
  tone control, 101
Address flow, 246
Addresses, 233
Addressing
  pattern, 233
  system, 231
Aging, 3
All-pass filter, 101
Alpha cutoff frequency, 162
Amplifier
  audio, 77, 212
  bias stabilization, 109
  characteristics, 81, 87
  class A, AB, C, 212
  complementary symmetry, 213, 219
  current, 83
  driver, 213

  operational, 6
  preamplifier, 79
  RF, 135
  voltage, 83
Analog computer, 15
Analysis, real-time, 231
Analyzer, logic-state, 232
Angle, phase, 33
Audio mixer, 28
Autotransformer coupling, 144

Backswing, 203
Balanced
  circuit, 6
  output, 216
Band
  elimination filter, 27
  width, 139
    power, 77
Bandpass
  filter, 27, 43
  value, 139
Base-spreading resistance, 83, 154
Bead, ferrite, 228
Beta, 80
  cutoff frequency, 162

270 · Index

Bias
  circuit, 113
  forward, 87
  signal-developed, 172
  stabilization, 109
  techniques, 117
Bipolar transistor, 79
Bistable multivibrator, 190
Bits, 237
Blocking oscillator, 201
Bogie values, 2, 5, 29
Breadboard model, 49
Breakdown
  diode, 130
  voltage, 5
Bridge
  balance, 223
  configuration, 222
  unilateralization, 155
Buffer, 176
Buffering, 231
Burn-in, 22
Bus lines, 237

Calculated risk, 11
Calculator-aided design, 157
Capacitance coupling, 145
Capacitor color code, 256
Central processing unit, 243
Ceramic capacitor, 229
Choke, rf, 228
Clamp diode, 195
Clapp oscillator, 177
Class A, B, C, 229
Clipping distortion, 98
Clock, 17, 18
  signals, 244
Color code, resistor, 254
Command
  bits, 244
  bus, 244
Comparative device parameters, 86
Complementary symmetry, 6
Component tolerances, 29
Compound connection, 220
Compression distortion, 98

Computer-aided design, 15
Conditional validity, 13
Confidence
  evaluation, 23
  factor, 23
  interval, 23
  level, 23
  limits, 23
Colpitts oscillator, 183
Conservative design, 166
Coupling circuit, 27
  coefficient, 148
  critical, 148
CPU, 246
CROM, 244, 246
Crossover distortion, 226
Crystal oscillator, 180
Current gain, 5
  variation, 221
Cursor, 252
Cutoff
  distortion, 6
  frequency, 5, 28

Darlington connection, 226
Data
  dependent errors, 238
  domain, 240
  flow, 231
  inputs, 245
  lines, 250
  multiplexer, 244
  readings, 238
  storage register, 159
  stream, 237
  transfer, 237, 246
dB, 29
dBm, 165
Debugging, 22, 231
Decade, 35
Decibel, 29
Decoupling circuit, 27
Degeneration, 226
Delay time, 189
Derating factor, 19
Design, conservative, 166

Index · 271

Device tolerances, 51
Diagram, flow, 222
Differential, 37
Differentiation, 37
Digital computers, 15
Diode
  clamping, 195
  steering, 194
Display
  interpretation, 246
  map, 237
Distortion
  crossover, 226
  cutoff, 6
  offset, 224
  stretching, 226
Distribution, normal, 11
Double
  humped response, 150
  tuned network, 146
  tuned transformer, 150
Dropout, oscillator, 164
Dual clock, 237, 243

Edge transition, 18
Effect, flywheel, 171
Effective resistance, 39
Emitter degeneration, 154
Environmental
  conditions, 6
  testing, 21
Equivalence, 3
Equivalent circuit, 42, 141
Equalizer, 151
Errors, dependent, 238
Exclusive OR, 233

Factor, derating, 19, 38
Fail-safe, 21
Fall time, 189
Feed
  series, 227
  shunt, 228

Feedback loop, 17
Feedthrough, 230
Ferrite bead, 228
FET
  input resistance, 87
  output resistance, 87
  power gain, 87
  transconductance, 87
  voltage gain, 87
Filter
  band elimination, 45
  bandpass, 43
  characteristic impedance, 62
  composite, 59
  constant-k, 58
  high-pass, 30
  L section, 55
  low-pass, 23, 30
  m-derived, 59
  multisection, 65
  parallel-T, 46
  pi-section, 56
  RC, 35, 47
  wave, 51
Flag
  signals, 240
  status, 246
Flip-flop, 190
Flow
  data, 231
  diagram, 222
  parametric, 243
Flywheel effect, 171
Forbidden region, 6
Four-phase clock, 244
Frequency
  chart, 30
  jump, 185
  multiplier, 227
  stability, 162, 165
Functional analysis, 243

Gain equalization, 151
Gates, logic, 265
General bias circuit, 113
Generator resistance, 83

272 · Index

g-force, 22
Glitch, 18, 237
Graphical solutions, 41, 89, 263
GT-cut crystal, 180

Hardware, 231
Harmonic
   second, 139
   third, 139
Hartley oscillator, 178
Heterologous tolerances, 4
High
   current multivibrator, 188
   cut filter, 27
   fidelity, 77
   pass filter, 27
Homologous tolerances, 4
Hot chassis, 20

Ideal stabilization, 21
IMP, 246
Impedance
   characteristic, 52
   matching, 150
   ratios, 140
Incoming inspection, 76
Incremental resistance, 79
Indication accuracy, 4
Induced voltage, 205
Infinite ladder, 52
Input
   data, 245
   impedance, 32, 79, 139
   port, 78
   resultant, 8
Instability, oscillator, 162
Integrated microprocessor, 246
Integration, 37
Interaction, 43
Interface, 246
Interlock, 20
Internal feedback, 153
Interstage
   coupling, 147
   network, 143

JFET, 173
JK flip-flop, 16
Jump, frequency, 185
Junction
   capacitance, 142
   diode surge protection, 133
   field effect transistor, 173

Kirchhoff's law, 33
kQ product, 148

Ladder, infinite, 52
Large-signal operation, 87
Latch-up, 76
Life test, 22
LIFO, 244
Limitation, frequency, 107
Lines
   data, 250
   termination, 55
Load
   filter, 56
   resistance, 82
Logic
   debugging, 22, 231
   state, 231
      analyzer, 233
      flow, 243
   systems, 231
Loop
   address, 247
   problem, 241
Loss, insertion, 101
Low-current multivibrator, 189
Low-cut filter, 27
Low-pass filter, 27
L section, 52

Manufacturer's rating, 107
Map
   display, 237
   pattern, 235
Matched pairs, 6

Mathematical
  calculator, 157
  model, 12
Maximum
  power gain, 80
  usable gain, 77
M-derived filter, 59
Mechanical tolerance, 4
Microphone, 78
Microprocessor, 231
Minority carriers, 203
Mixer, audio, 28
Model, prototype, 16
Monte Carlo method, 49
MPG, 80
mtbf, 23
MUG, 77
Multiplexer
  data, 244
  time, 246
Multiplier, frequency, 227
Multivibrator
  astable, 186
  bistable, 190
  monostable, 197
  nonsaturating, 190
  saturating, 190
  Schmitt, 199
  triggered, 190
Mutual inductance, 148

NAND
  function, 242
  gate, 241
Negative-edge
  triggering, 16
  feedback, 6, 91
Network, 3
  double-tuned, 146
  impedance, 142
  neutralization, 152
  neutralizing capacitor, 229
  unilateralization, 152
Noise, 25, 237
  factor, 226
Nominal values, 2

Nondescriptive answer, 13
Nonphysical answer, 13
Nonsinusoidal oscillators, 186
Normal distribution, 11
Normalization, 37
NPO, 168
Null, 45

Octave, 35
Offset distortion, 224
Ones-catching FF, 16
Op amp, 9
Operating
  efficiency, 177
  mode, 179
Operation code, 247
Operational
  amplifier, 6
  stack, 158
Operations, production, 1
Optimum Q, 148
Oscillation, parasitic, 228
Oscillator
  blocking, 201
  circuit design, 161
  Clapp, 171
  Colpitts, 183
  crystal, 182
  dropout, 163, 164, 183
  Hartley, 178
  instability, 162
  nonsinusoidal, 186
  piezoelectric, 182
  triggered, 201
  variable frequency, 173, 178
Output
  balanced, 216
  impedance, 32
  unbalanced, 216
  voltage, 34
Overtone
  crystal, 185
  frequency, 184

Paging technique, 246
Pairs, matched, 6

Parametric flow, 243
Parasitic oscillation, 160, 228
Partial emitter degeneration, 154
Pattern
  addressing, 233
  map, 235
Peripheral devices, 245
Phase angle, 33
Physical property, 12
Pocket calculator, 15
Power
  capability, 10
  dissipation, 10
  efficiency, 136
  transfer, 136
ppm, 136
Primary, tapped, 144
Probability, 11
  curve, 11
Probe connections, 246
Problem, loop, 241
Production
  lot, 1
  operations, 1
  tests, 23
Program
  deviations, 252
  sequences, 238
  sequencing, 244
Prototype model, 16, 23
Punch-through, 6
Push-pull
  operation, 162
  oscillator, 179

Q, 137
Q multiplication, 165
Quadratic form, 14
Qualifier, 246
Quality, 23
  assurance, 24
  control, 24, 49
  engineering, 24
  factor, 137
Quartz crystal, 182

Radiation, 21
RALU, 244, 246
Rating, power, 217
Ratio, signal/noise, 226
RC
  filter, 35, 47
  time-constant chart, 48
Real-time analysis, 231, 243
Regenerative feedback, 179
Region
  active, 206
  cutoff, 206
  saturation, 206
Register
  shift, 244
  storage, 159
Registration, 238
Relaxation oscillator, 186, 210
Resistance
  characteristic, 52
  thermal, 5
Resistive tolerances, 4
Resistor color code, 254
Resonant circuits, 137
Response curves, 151
  double-humped, 150
Resultant input, 8
Ringing, 237
Rise time, 188
Rolloff, 34
ROM, 231
Room, screened, 160

Screen room, 160
Second harmonic, 139
Sequences, program, 238
Sequencing, program, 244
Series feed, 227
Set/reset, 245
Shift register, 244
Shock, 25
Shunt feed, 228
Sign extension, 245
Significant feedback, 94

Index · 275

Signal
   developed bias, 172
   to noise ratio, 226
Signals, flag, 240
Simulation, 235
Single-shot, 238
Small-signal operation, 80
Speed specification, 3
Spurious answer, 13
Stability, frequency, 162
Stack, operational, 158
Stagger tuning, 150
State
   logic, 232
   time, 240
Status flag, 246
Steering
   negative pulse, 194
   positive pulse, 195
   trigger pulse, 191
Storage time, 189
Stretching distortion, 226
Swamping resistor, 229

Tantalum capacitor, 229
Tapers, resistance, 269
Tapped primary, 144
Temperature
   characteristics, 167
   coefficient, 26
   compensating capacitor, 167
Thermal resistance, 5
Third harmonic, 139
Time
   constant, 36
   delay, 189
   domain, 240
   multiplex, 246
   real, 243
   state, 240
   storage, 189
Tolerances
   component, 29
   device, 5
Torsion, 1, 25

T pad, 3
Transfer
   data, 237
   power, 136
Transformer
   action, 141
   coupling, 142
   double-tuned, 150
Transient
   response, 35
   state, 206
Transistor
   current gain, 81, 85
   input resistance, 81, 82
   output resistance, 81, 83
   power gain, 81
   voltage gain, 81, 84
Trigger
   pulse steering, 191
   word, 238, 248
Triggered blocking oscillator, 201
Truth table, 241
TTL, 244
Tuned primary, 142
Tuning, stagger, 150
Tunnel diode, 14

UJT oscillator, 210
UL, 20
Unbalanced output, 216
Underwriters Laboratories, 20
Unilateralization, 152
Universal time-constant chart, 148

Validity, conditional, 13
Values, bogie, 2, 5, 29
Variable-frequency oscillator, 173, 178
VFO, 173, 178
Vibration, 25
Virtual ground, 9
Voltage
   feedback, 94
   induced, 205

Voltage *(Continued)*
   operated device, 85
   stabilized amplifier, 132

Wave
   filter, 51
   trap, 68
Word generator, 233
Worst case, 3, 29

X-cut crystal, 180
XOR, 234

Zener diode, 131
Zero
   frequency, 52
   voltage point, 223